STP 1451

Roofing Research and Standards Development: 5th Volume

Walter J. Rossiter, Jr. and Thomas J. Wallace, editors

ASTM Stock Number: STP1451

ASTM International
100 Barr Harbor Drive
PO Box C700
West Conshohocken, PA 19428-2959

Printed in the U.S.A.

ISBN: 0-8031-3469-X
ISSN: 1050-8104

Peer Review Policy

Each paper published in this volume was evaluated by two peer reviewers and at least one editor. The authors addressed all of the reviewers' comments to the satisfaction of both the technical editor(s) and the ASTM International Committee on Publications.

To make technical information available as quickly as possible, the peer-reviewed papers in this publication were prepared "camera-ready" as submitted by the authors.

The quality of the papers in this publication reflects not only the obvious efforts of the authors and the technical editor(s), but also the work of the peer reviewers. In keeping with long-standing publication practices, ASTM International maintains the anonymity of the peer reviewers. The ASTM International Committee on Publications acknowledges with appreciation their dedication and contribution of time and effort on behalf of ASTM International.

Printed in Bridgeport, NJ
August 2003

Foreword

The Symposium on Roofing Research and Standards Development was held in Tampa, Florida on December 7, 2003. ASTM International Committee D08 on Roofing and Waterproofing served as sponsor. Symposium chairmen and co-editors of this publication were Thomas J. Wallace, Coast to Coast Consultants, Port Richey, Florida; and Walter J. Rossiter, Jr., National Institute of Standards and Technology, Gaithersburg, Maryland.

Contents

Overview

Sound standards have strong technical bases. This symposium on *Roofing Research and Standards Development* is the fifth in a series dating back to the mid-1980s. It was conducted under the auspices of Subcommittee D08.21 on Research Needs for Roofing and Waterproofing. This symposium and the papers described in the Proceedings illustrate D08's commitment to developing standards that have strong technical bases, which ultimately contributes to improve roofing performance. Proceedings in this series are: *Roofing Research and Standards Development, ASTM STP 959* (1986), edited by R.A. Critchell, *Roofing Research and Standards Development, 2nd Volume, ASTM STP 1088* (1990), *Roofing Research and Standards Development, 3rd Volume, ASTM STP 1224* (1994), *Roofing Research and Standards Development, 4th Volume, ASTM STP 1349* (1998), all edited by T.J. Wallace and W.J. Rossiter, Jr.

When this *Roofing Research and Standards Development* series was initiated in 1986, the roofing industry was in a state of rapid flux. For example, many new membrane products for commercial and industrial roofs—particularly those based on elastomers, thermoplastics, and polymer-modified bitumens—appeared on the market without adequate characterization that they were suitable to withstand the rigors of the roof environment. Consequently, in some cases, these new membrane products disappeared from the market almost as quickly as they arrived. That era is long behind the industry as fewer new products were introduced during the 1990s. That is, the industry settled into a period of stability during the 1990s, which has now extended into the new millennium

When the industry was in a rapid state of flux, D08 was under extreme pressure to develop materials specifications for the new products. Its members responded admirably and the needed specifications were issued after considerable debate and discussion. With stability settling in, D08 turned its attention to developing and improving companion standards such as test methods for evaluating important performance properties, and practices for inspection, maintenance, and application.

In essence, the stabilization of the industry has provided opportunity for D08 to address long-standing issues, and increase its understanding of factors that vitally affect successful long-term performance. Paramount among these factors are characterization of durability, wind resistance, fire resistance, and impact on building energy consumption. The symposium papers assembled in these current Proceedings directly target these four factors. They represent an important contribution to D08's commitment to expanding the knowledge base that supports successful roof performance. The papers are also exemplary of the types of studies that will be needed continuously to improve roofing performance. Five papers address durability and material characterization, two focus on wind resistance, two apply to energy consumption, and one centers on fire resistance. Consistent with the title of the symposium series, in many cases, the authors have made recommendations for development of new ASTM standards or improvement of those already issued. The editors of these Proceedings hope that the D08 members will review, digest, and critique these recommendations and, as appropriate, initiate task group activities to consider them in the D08 standards development process.

As in the past, these Proceedings are dedicated to the members of ASTM Committee D08 who unselfishly give their time and energy to improve the performance of roof systems. The editors express their sincere thanks and appreciation to those individuals who participated in the organization and conduct of the symposium. D. M. Bailey, C. G. Cash, R. M. Dupuis, M. F. Franks, M. S. Graham, J. R. Hagan, W. A. Kirn, M. Knowles, L. R. Meyers, R. M. Paroli, G. A. Smith, T. L. Smith, R. J. Wallace, and J. R. Wells were members of the steering committee. At ASTM headquarters, Dorothy Fitzpatrick, Joseph Hugo, and Pat Picariello organized the symposium arrangements. Maria Langiewicz, working with Crystal Kemp, directed the review and publication of the papers. The editors acknowledge with thanks the noteworthy contributions of Ms. Langiewicz for her diligence and thoroughness in keeping us on schedule as these Proceedings were developed. R. J. Goode was the ASTM Committee on Publications (COP) representative to the symposium steering committee. Above all, specials thanks are given to the authors and reviewers of the ten papers without whose outstanding efforts in writing and reviewing, respectively, and commitment to remaining on schedule, the symposium and Proceedings would not have been possible.

Thomas J. Wallace
Coast to Coast Consultants
Port Richey, FL 34668
Symposium Co-chair and Editor

Walter J. Rossiter, Jr.
National Institute of Standards and Technology
Gaithersburg, MD 20899-8615
Symposium Co-chair and Editor

Session I: Uplift and Fire

Ming L. Shiao,[1] *Richard A. Snyder,*[2] *Robert D. Livsey,*[2] *and Husnu M. Kalkanoglu*[2]

Measuring Uplift Resistance of Asphalt Shingles

REFERENCE: Shiao, M. L., Snyder, R. A., Livsey, R. D., and Kalkanoglu, H. M., "**Measuring Uplift Resistance of Asphalt Shingles,**" *Roofing Research and Standards Development: 5th Volume, ASTM STP 1451*, W. J. Rossiter and T. J. Wallace, Eds., ASTM International, West Conshohocken, PA, 2003.

ABSTRACT: The uplift resistance of asphalt shingles with self-sealing adhesives is studied by applying mechanical forces to examine the effects of loading geometry, shingle rigidity, and the rate of load application. Three distinct test methods for simulating the wind-induced uplift effects of peeling, pulling, and its combination are used to study the sealant performance as a function of shingle rigidity and the rate of loading. The results show that the measured sealant strength is significantly influenced by the test and loading configurations, as well as by the loading rate. The current ASTM D 6381 test, which measures the uplift resistance under mechanical peeling action, is found to yield relatively low measured sealant strength as compared to the other test method intended to simulate wind uplift effects according to current wind-uplift force model. Fracture analysis and fractography conducted on test specimens reveal that the measured shingle uplift resistance is significantly increased under high deformation rate due to a transition from failure inside the asphalt coating to adhesive fracture in the sealant by cavitation mechanism.

KEYWORDS: asphalt shingle, sealant, wind load, uplift resistance, mechanical test, adhesion

Introduction

The wind resistance of various building materials, including asphalt shingles, has received increasing attention recently due to several major windstorms that have resulted in costly damages [*1-2*]. As a result, there have been increasing proposals for change in wind-load design or building codes [*3-5*] and new developments for test methods to evaluate the wind resistance of shingles [*6*]. In response to such a trend and the lack of scientific understanding on the wind uplift mechanism for asphalt shingles [*7*], the Asphalt Roofing Manufacturer's Association (ARMA) in collaboration with researchers at Colorado State University established a wind research program to study the major wind uplift mechanisms for self-sealing asphalt shingles [*7,8*]. They have successfully developed a model for wind uplift in asphalt shingles [*9*] and have also experimentally derived the magnitude of the wind uplift forces acting on the both sides of sealant strip for various styles of shingles [*8*]. This wind load model not only provides a clearer picture of how asphalt shingles are loaded under high wind conditions, but also opens the

[1] Principal Research Engineer, Exterior Products Group, CertainTeed Corp., 1400 Union Meeting Rd., Blue Bell, PA

[2] Manager-Product Compliance, Product Development Engineer, and Director of RPG Technology, respectively, CertainTeed Corp., 1400 Union Meeting Rd., Blue Bell, PA

opportunity for developing new test methods that may utilize mechanical forces to evaluate the resistance of shingles to such wind-induced forces.

While the magnitude of the wind-induced uplift forces has been well described by the new wind-load model [*9*], the evaluation of a shingle's resistance to such loading through mechanical means may not be an easy subject. It is well known that the behavior of complex viscoelastic adhesives like asphalt sealant is difficult to quantify. For example, their deformation and fracture may involve dynamic phenomena such as meniscus instability, cavitation, and even the breakdown of fibrils [*10-14*]. Furthermore, the strength of an adhesive, or its stickiness, can sometimes be quite paradoxical. One of the most striking is that the adhesive may not be sticky by itself: its thickness and the mechanical and surface properties of the probe used to test its apparent strength can be important [*10*]. For instance, if you get stuck when walking on an asphalt pavement that has softened on a horribly hot summer afternoon, you'll have a better chance of getting away if your shoes have flexible soles. Such an example clearly demonstrates that the physics of peeling and of pulling off are definitely different and thus need to be carefully examined. This is especially true when evaluating asphalt sealants under various loading geometries, shingle rigidities, shingle design complexities, and the rates of deformation.

In this study, the uplift resistance of asphalt shingles with factory-applied self-sealing adhesives is examined by applying mechanical forces in different loading modes for peeling, pulling, or a combined action. Three distinct test methods are used to study the performance of the sealant as influenced by test configurations, shingle rigidity, and by rate of deformation. The ASTM Test Method for Measurement of Asphalt Shingle Tab Mechanical Uplift Resistance (D 6381) and a Direct-Uplift Sealant Test method can be regarded as the two extremes of adhesive testing where one is dominated by a peeling force and the other completely by tensile stress. The third test method examines the shingle uplift resistance due to wind forces acting simultaneously on both sides of the sealant as suggested by the wind load model developed by Perterka et al. [9].

Test Methods for Measuring Shingle Uplift Resistance

ASTM Test Method for Measurement of Asphalt Shingle Tab Mechanical Uplift Resistance (D 6381)

This test method is designed to measure the sealant's uplift resistance by mechanically grabbing and lifting shingle at the windward side of the sealant. Under such test configuration, the loading point and the sealant strip will not be in the same plane for most cases and the sealant is therefore considered to be stressed largely by peeling action due to the resultant shear stresses. Also, since the shingle dimension on the windward side of sealant is relatively short, the peeling angle with respect to sealant centerline will thus vary significantly during the test. As a result, the apparent sealant strength as measured by this test will be influenced by factors such as shingle design, sealant position, and shingle rigidity due to the fact that the angle of peeling is known to affect the adhesive strength during a peeling test [*15*]. Hence, the test may not necessarily reflect the true sealant performance. In particular, one should be cautious when the resistance value obtained by this test is directly compared to those uplift forces as calculated from Perterka's wind-load model [*9*], since these two cases represent different loading geometry with respect to sealant performance.

Direct-Uplift Sealant Test or the "T-Pull" Test

This test method is similar to that of the probe-tack test for adhesives [*10*], except that an actual shingle is used as the surface of the probe. In this case, the sealant is tested under direct tension as the probe, or the actual sealed shingle, is lifted vertically. This is shown in Fig. 1, Fig. 2A, and Fig. 2B, where the test specimen consists of a 102 mm × 121 mm (4 in. × 4¾ in.) bottom shingle piece with sealant strip in the middle and a top piece of shingle of 95 mm × 38 mm (3¾ in. × 1½ in.) sealed to the sealant strip at the bottom shingle. An aluminum "T" section of the same dimension as the top piece of shingle is then solidly glued by epoxy to the top piece of shingle and used as the application point for the uplift force. The test fixture, as shown in Fig. 2A, is similar to that used in ASTM D3462 for measuring the fastener pull-through resistance, except that a 3 mm (1/8 in.) thick aluminum mask of size 102 mm × 121 mm (4 in. × 4¾ in.) with a central opening of 45 mm × 102 mm (1¾ in. × 4 in.) is used to restrain the bottom piece of shingle during uplift testing. An example of the "T-Pull" test in progress is shown in Fig. 2B.

Wind-Model Shingle Uplift Test or the "Double-Peel" Test

This test method was specifically developed to test the sealant performance with the wind uplift force distribution as derived from the Perterka's wind load model [*9*]. In his

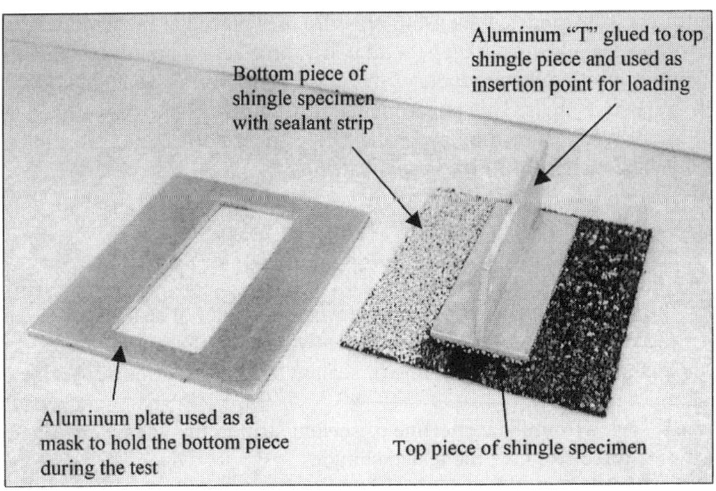

FIG. 1 – *The test specimen configuration for the Direct-Uplift Sealant Test, or the "T-Pull" Test.*

FIG. 2A & 2B – *The test fixture and actual testing using the T-Pull Test.*

model, the uplift forces are found to result from the positive pressure by wind impinging upon the shingle front edge and a local negative pressure from air flow accelerated over the upper shingle surface. In subsequent experimental wind-tunnel studies on various types of commercially available shingles, the wind uplift forces were determined through measuring uplift pressure coefficients using pressure taps placed at the windward and leeward sides of the sealant strip [*8*]. The total uplift force acting on the sealant strip was then calculated by summing the products of the uplift pressures on each side of sealant strip and their respective areas. This experimental procedure has been developed as a standard ANSI-approved test method by Underwriter Laboratories, Inc., Northbrook, IL, named UL 2390. The total uplift force is calculated as

$$F = F_1 + F_2 = V^2 \cdot DC_{p1} \cdot L_1 \cdot K_a \cdot K_b + V^2 \cdot DC_{p2} \cdot L_2 / 2 \cdot K_a \cdot K_b \quad (1)$$

where V = the velocity of the wind in mph
DC_{p1} = the wind uplift coefficient on the windward side of sealant
DC_{p2} = the wind uplift coefficient on the leeward side of sealant
L_1 = the distance from the centerline of sealant strip to the windward edge of shingle
L_2 = the distance from the centerline of sealant strip to the windward edge of the shingle directly above the tested shingle
K_a = a coefficient accounting for building height, exposure, terrain, etc.
K_b = a constant that converts the wind speed to a strength design load for the 95.3 mm (3¾ in.) specimen size

This can be visualized in Fig. 3, where the wind uplift forces acting on both sides of sealant strip account for the total uplift force *F*. Some values of the uplift forces calculated from Eq. (1) using those experimental wind uplift coefficients determined by Perterka et al [*8*] are listed in Table 1 for various styles of commercially available shingles. Note that the uplift force generated by the negative pressure on the leeward side of sealant can be significant, ranging from 30% to almost 130% of the uplift force on windward side of the sealant. As a result, the performance of the sealant would need to be properly evaluated by mechanical loading acting simultaneously on both sides of the sealant. Also, the uplift forces in a laminate shingle are found to be similar between the one and two layer portions of the tab, whereas the rigidity of the shingle is approximately double in the two-layered portion of the tab [*8*]. Under such circumstance, the rigidity of such shingle may have significant impact on the behavior of the sealant, particularly when the uplift forces acting in front of and behind the sealant are considered.

In order to understand the performance of the sealant strip when subjected to wind uplift forces acting simultaneously on both sides of the sealant, the Double-Peel Test method was developed by constructing a text fixture that cleates mechanical forces in proportion to the experimentally measured F_1 , F_2 values to be applied simultaneously on both sides of the sealant strip. This is shown in Fig. 4, where an aluminum test fixture with two adjustable specimen clamps for applying mechanical uplift forces can be seen. The relative magnitude of the uplift forces F_1 , F_2 can be adjusted to achieve the experimentally measured values by adjusting the position of the total uplift force *F* relative to the uplift force F_1. This can be done by calculating the distance *d* in Fig. 4 using the balance of moments with respect to the application point of the uplift force F_1

$$d = \frac{F_2}{F_1 + F_2} \cdot \frac{L_2 + 2L_1}{4} \tag{2}$$

In Eq. (2), the distance *d* is directly proportional to the magnitude of the uplift forces acting on the leeward side of the sealant. For single-layer strip shingle, the distance *d* is a relatively small dimension indicating that most of the uplift forces are on the windward side of the sealant. However, in some multi-layered or complex shingles the distance *d* can be quite significant and may approach the midpoint between F_1 and F_2, thus giving rise to the "double peeling" effect. The test fixture shown in Fig. 4 is designed with adjustable specimen clamps to place the uplift forces along the centerline of each uplift area for various shingle styles and an adjustable swivel head for locating the total uplift force *F* according to the distance *d*. Thus, the sealant performance under the uplift forces acting on both sides of sealant can be evaluated using such test fixture by properly setting the positions of the specimen clamps and the distance of the swivel head to reflect the application points for the uplift forces F_1 , F_2 , and the total uplift force *F*.

The shingle specimen for the double-peel test consists of a top piece of shingle with size 95 mm × 127 mm (3¾ in. × 5 in.) and a bottom piece of shingle 38 mm × 95 mm (1½ in. × ¾ in.) containing the sealant strip to be tested. The bottom piece of shingle with the sealant strip is first sealed to the top piece according to the sealing condition as described in the next section, and then an aluminum "T" section of the same dimension as the bottom piece of shingle was glued to the top piece shingle by epoxy. This is shown in Fig. 5 for the setup of a test specimen during the double-peel test.

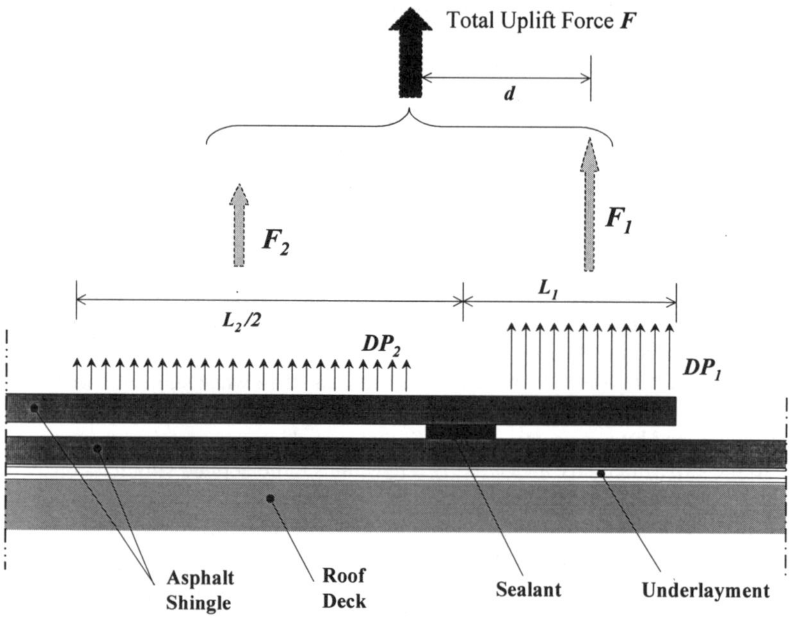

FIG. 3 – *The schema of the wind uplift forces from the model by Perterka [9].*

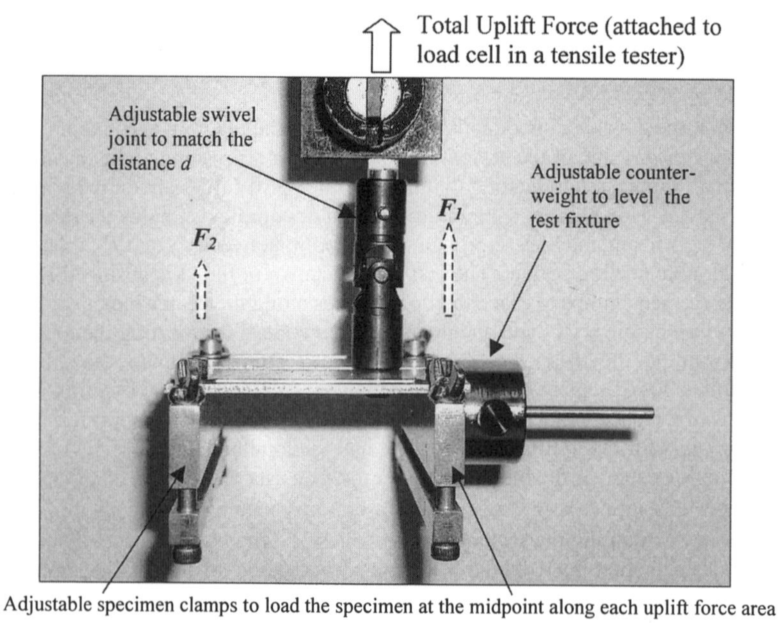

FIG. 4 – *Test fixture for the Wind-Model Shingle Uplift Test or the Double-Peel Test.*

FIG. 5 – *An example of the Wind-Model Shingle Uplift in progress.*

TABLE 1 – *Some values of the uplift forces calculated from the data by Perterka et al [8] using category E wind (150 mph).*

Shingle Type	Uplift Force F_1, kg	Uplift Force F_2, kg	Total Uplift Force, F	$\frac{F_2}{F_1}$
Light-weight fiberglass strip shingle	-1.13	-0.86	-1.99	0.76
Light-weight fiberglass strip shingle	-1.33	-0.39	-1.72	0.30
Light-weight fiberglass strip shingle	-1.25	-0.48	-1.72	0.38
Mid-weight fiberglass strip shingle	-0.78	-0.80	-1.58	1.02
ASTM Type I organic strip shingle*	-2.00	-1.49	-3.49	0.75
Heavy-weight fiberglass strip shingle	-0.58	-0.36	-0.94	0.62
Mid-weight fiberglass laminate shingle - one layer	-0.95	-0.22	-1.17	0.23
Mid-weight fiberglass laminate shingle - two layer	-1.09	-0.30	-1.39	0.27
Heavy-weight fiberglass laminate shingle*	-1.78	-2.28	-4.06	1.28

*: The uplift forces were averaged across their exposure tabs.

Sample Information and Experimentation

To study the effect of shingle rigidity, two types of shingle were selected. The first shingle sample was a commercially available, mid-weight 3-tab fiberglass shingle that represents the case of a single-layer, flexible shingles. For increased shingle rigidity, a multi-layered, heavy-weight laminate fiberglass shingle was selected from the same shingle manufacturing facility. Both samples were collected within a close time frame of manufacturing in the hope that they may have a similar sealant composition, as well as similar shingle raw material composition. For the multi-layered shingle sample, the exposed tab is significantly larger in dimension and the sealant strip is located on the back of the shingle. Also, test specimens of the laminated shingle were cut only from the tabs with two layers of materials. The test specimens of the multi-layered shingle for the ASTM D6381 peel test had to be cut shorter than the actual shingle exposure length due to the limited size of the test fixture. This may have resulted in higher than actual values in the test, possibly due to decreased peeling action and/or change in peeling angle. All test specimens were cut from the shingle samples using a heavy-duty cutting board. The specimens were then sealed at 57.2±1°C (135°F) for 16 hours in an air-circulating dark oven. After removing from the oven and allowing to cool to room temperature, the specimens were either tested directly by the test methods or further prepared by securing the aluminum "T" section using low exotherm epoxy, cured overnight.

An Instron model 1125R universal tensile tester interfaced with a computer for data acquisition was used to conduct the tests. The sealed shingle test specimens were first randomized to form three subsets of specimens to be tested at loading rates of 12.7, 127, and 1270 mm/min. (0.5, 5, and 50 in./min.). For each specimen subset, at least ten specimens were tested. The test specimens were loaded to the point of failure and the maximum loads recorded. All shingle uplift resistance tests were carried out at room temperature. The fractured sealant in each specimen was carefully examined to judge the amount of adhesive/cohesive failure by estimating the amount of asphalt coating remaining on the failed sealant. Also, some representative specimens were collected and were examined by stereo optical microscopy and scanning electron microscopy to study the failure mechanism of the sealant.

Results and Discussions:

The shingle uplift resistances as measured by the three test methods are listed in Table 2, as a function of the test rates and shingle rigidity. The results are also plotted in Fig. 6 and Fig. 7 for comparison among each test method and in Fig. 8 as a function of loading rates. As expected, the ASTM D6381 peel test gave relatively low results for sealant strength while the T-pull test, which represents the case of more rigid shingle, resulted in higher values of uplift resistance. When the samples were tested by the double-peel test, which intends to simulate the effects of wind uplift forces as suggested by Perterka's model, the resultant shingle uplift resistance was found to be in between the values obtained by the ASTM peel test and by the T-pull test. In particular, the shingle uplift resistance as measured by the double-peel test can be significantly higher than that of the ASTM D6381 peel test, ranging from 40% to more than 200% larger values (see Table 3). In the case of two-layered shingle sample, the ASTM peel test may give higher than expected value, due to the fact that the particular shingle sample selected for this study has an unusually long exposure tab and the test specimens had to be cut shorter than actual exposure length in order to fit into the test fixture of the ASTM peel test. This results in reduced span for uplift resistance testing and may increase the rigidity of the test specimen thereby reducing the peeling action during the test. On the other hand, the double-peel test for the same shingle sample with two layers was conducted according to the exposure length specified by the manufacturer. Thus, the smaller difference seen in the case of the two-layered shingle sample between the ASTM peel test result and the double-peel test result (see Fig. 7) may not necessarily reflect the actual sealant performance as it is strongly influenced by the two test methods. Nevertheless, the results reveal that the ASTM D 6381 peel test may lead to very conservative sealant performance, i.e., low measured sealant strength, especially when the actual wind uplift forces acting simultaneously on both sides of the sealant are considered.

The effect of the loading rate shown in Fig. 8 indicates that the shingle uplift resistance significantly increases as the loading rate increases, ranging from 57% increase to almost three times higher as the test rate increased from 12.7 to 1270 mm/min. Also, the amount of increase is found to be more significant in the case of more rigid, two-layered shingle than in the case of one-layered flexible shingles. Hence, when sealants are subjected to sudden gust winds resulting in rapidly imposed uplift forces, the actual sealant performance may far exceed the uplift resistance measured in the laboratory using procedures with low deformation rate, particularly those values measured by the ASTM D 6381 peel test. Furthermore, examination of the fractured

TABLE 2 – *Shingle uplift resistance values as determined mechanically by the three test methods and expressed in units of kg [lb].*

Loading Rate	Single-Layer Shingle Sample			Two-Layer Shingle Sample		
	Method 1	Method 2	Method 3	Method 1	Method 2	Method 3
12.7 mm/min.	3.99 kg (0.68)* [8.80 lb]	11.91 kg (0.70) [26.26 lb]	8.52 kg (1.25) [18.79 lb]	2.44 kg (0.48) [5.38 lb]	7.05 kg (1.62) [15.54 lb]	3.68 kg (0.53) [8.12 lb]
127 mm/min.	6.14 (0.77) [13.55]	17.98 (1.79) [39.65]	10.34 (1.99) [22.79]	4.60 (0.88) [10.15]	11.41 (1.88) [25.17]	6.39 (0.86) [14.08]
1270 mm/min.	6.26 (1.65) [13.81]	23.16 (1.89) [51.07]	13.87 (2.93) [30.59]	7.21 (0.93) [15.91]	13.09 (1.76) [28.87]	7.09 (1.05) [15.64]

Method 1: The ASTM D6381 Test Method for Measurement of Asphalt Shingle Tab Mechanical Uplift Resistance
Method 2: The Direct-Uplift Sealant Test, or the "T-Pull" Test (See text for details)
Method 3: The Wind-Model Shingle Uplift Test, or the "Double-Peel" Test
*: The numbers in the parenthesis are the 95% confidence intervals.

TABLE 3 – *The Relative Ratio of the Shingle Uplift Resistance among the Three Methods.*

Extension Rate, mm/min.	Single-Layer Shingle Sample			Two-layer Shingle Sample		
	12.7	127	1270	12.7	127	1270
ASTM Peel / T-Pull Test	0.335	0.342	0.270	0.346	0.403	0.551
Double-Peel / T-Pull Test	0.715	0.575	0.599	0.522	0.559	0.542
Double-Peel Test / ASTM Peel	2.137	1.684	2.217	1.508	1.387	0.983

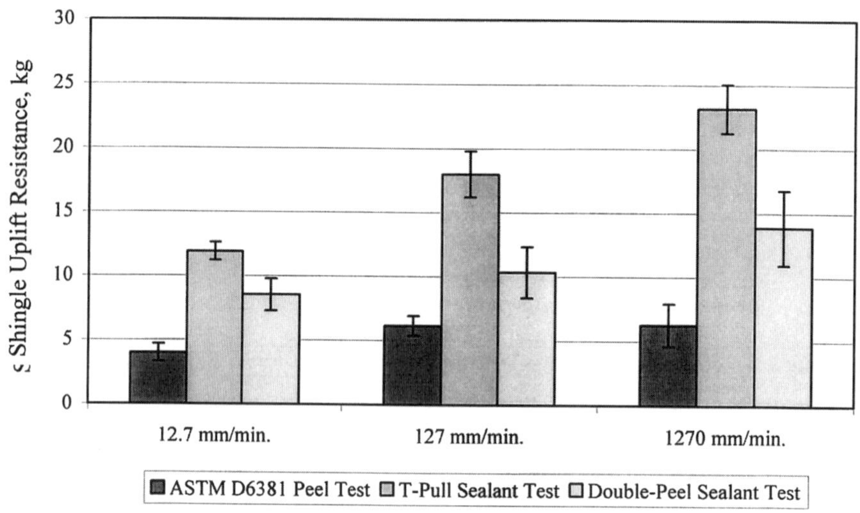

FIG. 6 – *The shingle uplift resistance measured by the three test methods for the one-layer shingle sample.*

sealants after each uplift test reveals that the failure mechanism changed from a more cohesive failure to a more typical adhesive failure as the test rate increases, regardless the type of shingles tested. This is shown in Fig. 9 to Fig. 11 for the three test rates applied on the one-layered shingle sample. In the case of the low testing rate at 12.7 mm/min., the sealant is found to fail in a cohesive manner by fracturing the asphalt coating either at the shingle's coating/glass mat interface or at the coating/back surfacer interface. For the T-pull sealant test, the failure is found mostly to occur at the asphalt/glass mat interface that leads to higher sealant strength as compared to the other two test methods, which are found to fail mostly at the coating/back surfacer and less at coating/glass mat interfaces

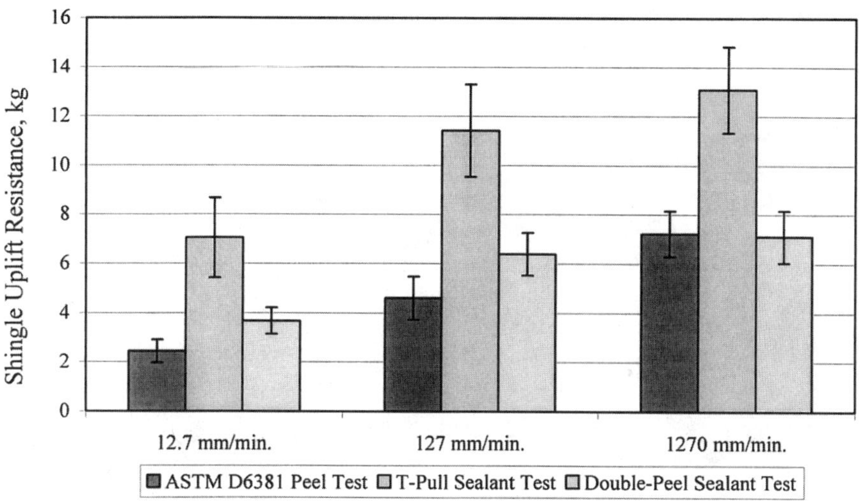

FIG. 7 – *The shingle uplift resistance measured by the three test methods for the two-layered shingle sample.*

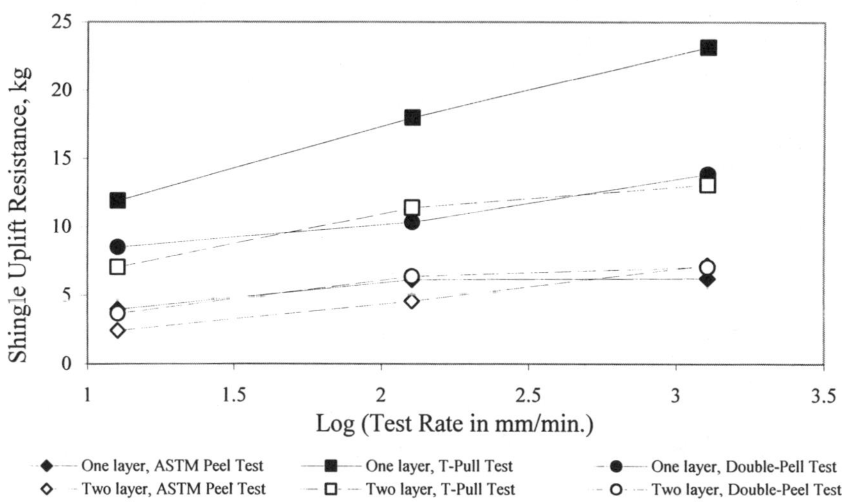

FIG. 8 – *The effects of loading rates on the shingle uplift resistance.*

due to increased shearing. As the test rate increases to 127 mm/min., the failure mode of the sealant appears to be of a mixed mode between the cohesive failure and the adhesive failure inside the sealant. Upon increasing the test rate up to 1270 mm/min., the fracture surfaces of the sealant become mostly adhesive failure due to fast fracturing through the initiation of cavitation under high triaxial stresses as the materials become more restrained by their viscoelastic nature. Such adhesive failure at high loading rate is found

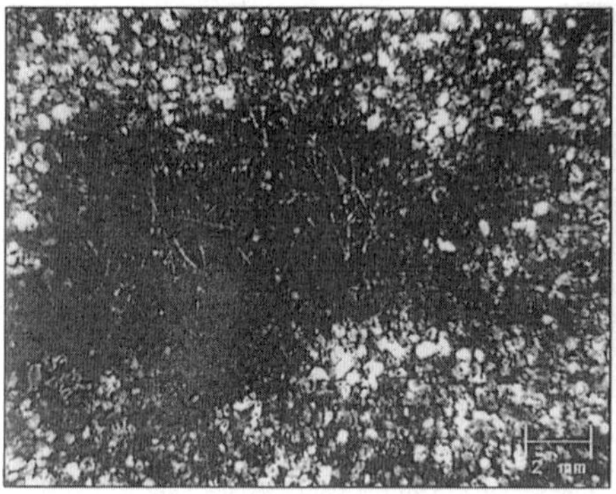

12.7 mm/min. – the back of top shingle

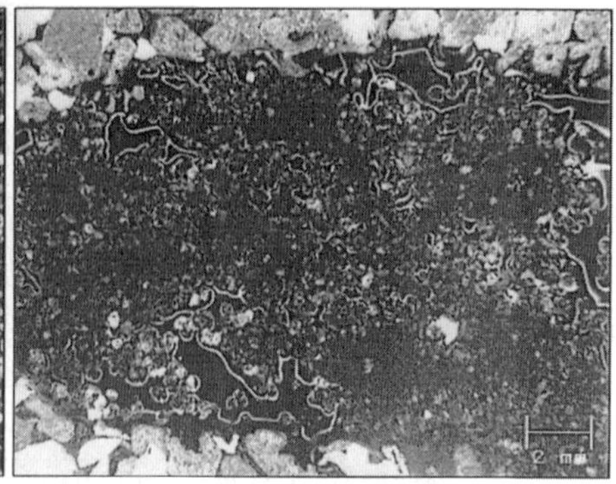

12.7 mm/min. – sealant at bottom shingle

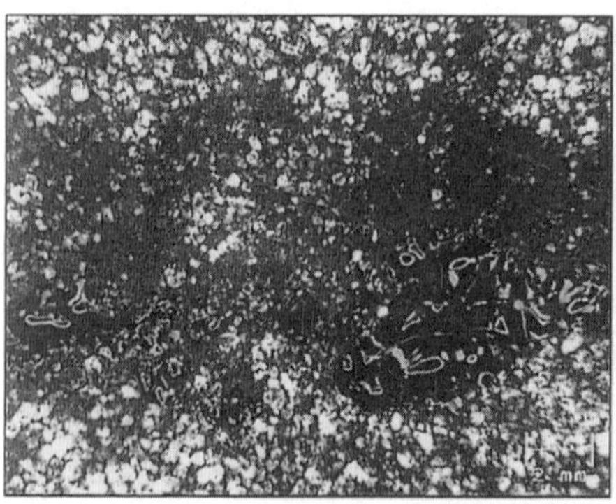

127 mm/min. – the back of top shingle

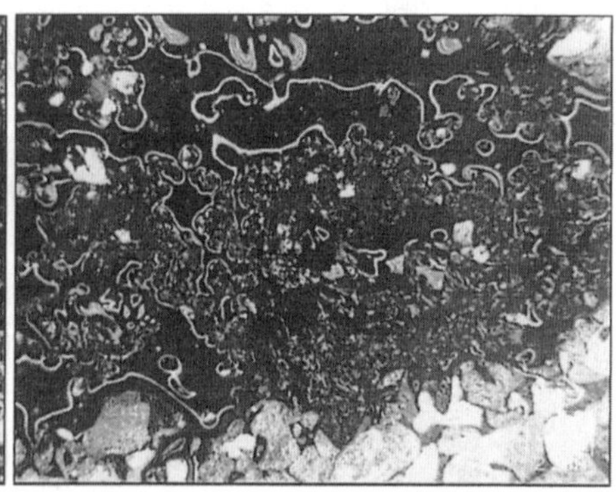

127 mm/min. – sealant at bottom shingle

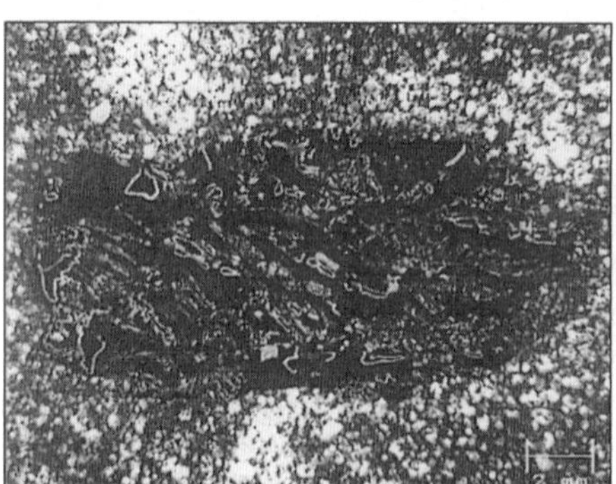

1270 mm/min. – the back of top shingle

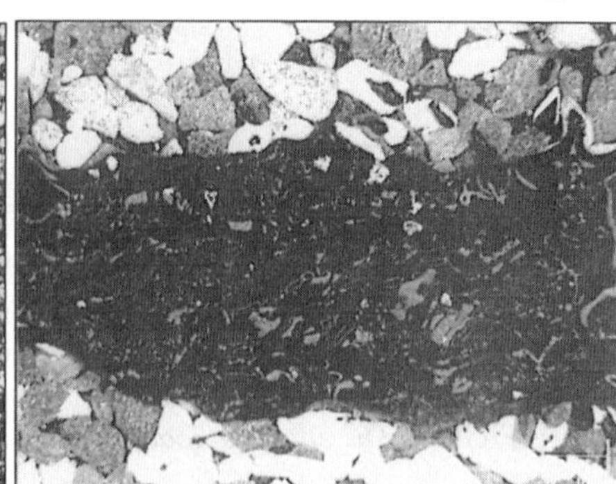

1270 mm/min. – sealant at bottom shingle

FIG. 9 – *The fracture surface of the sealant tested by the ASTM D 6381 Test.*

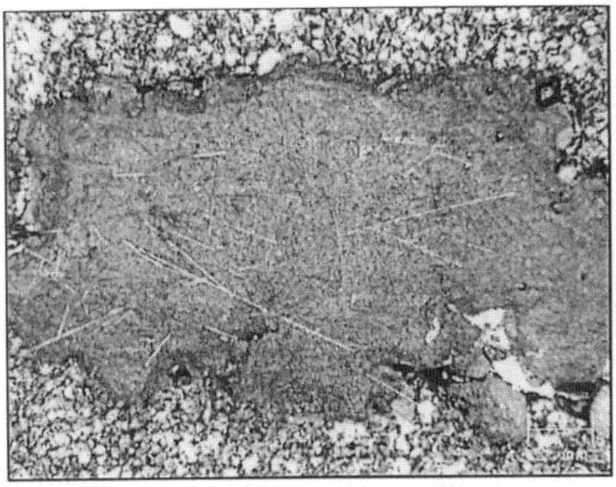

12.7 mm/min. – the back of top shingle

12.7 mm/min. – sealant at bottom shingle

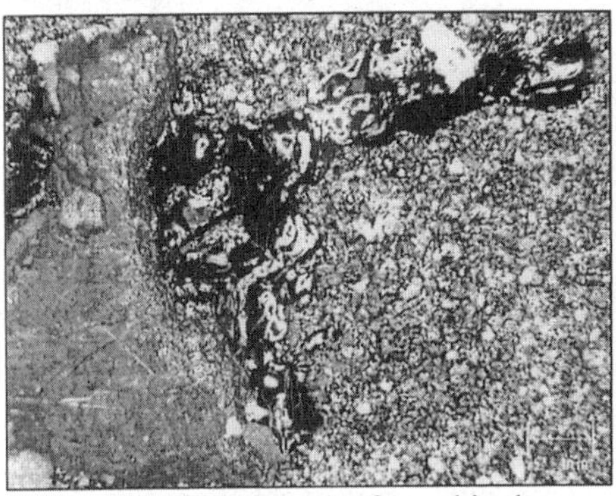

127 mm/min. – the back of top shingle

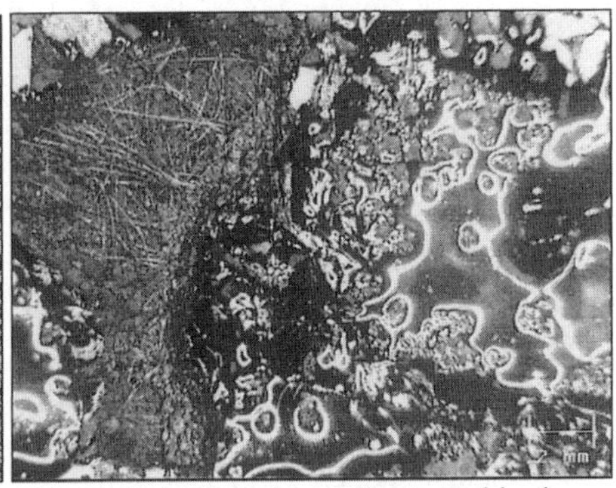

127 mm/min. – sealant at bottom shingle

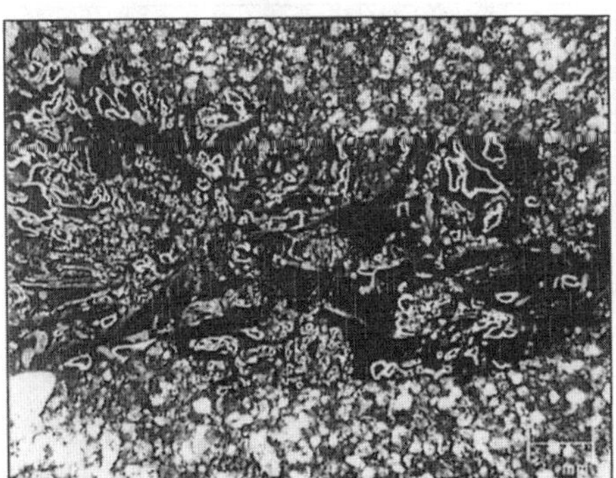

1270 mm/min. – the back of top shingle

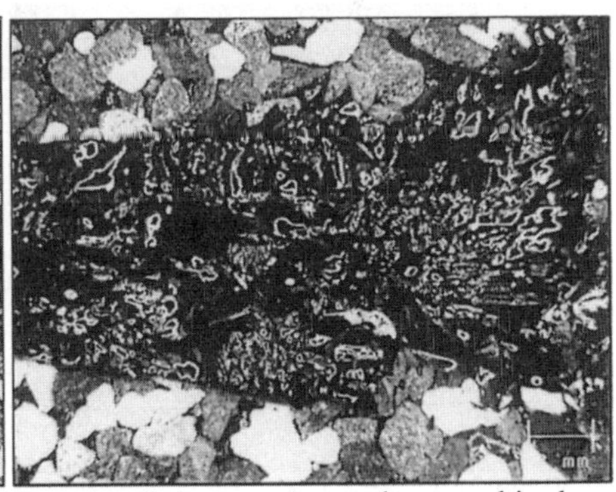

1270 mm/min. – sealant at bottom shingle

FIG. 10 – *The fracture surfaces of the sealant tested by the T-Pull Sealant Test.*

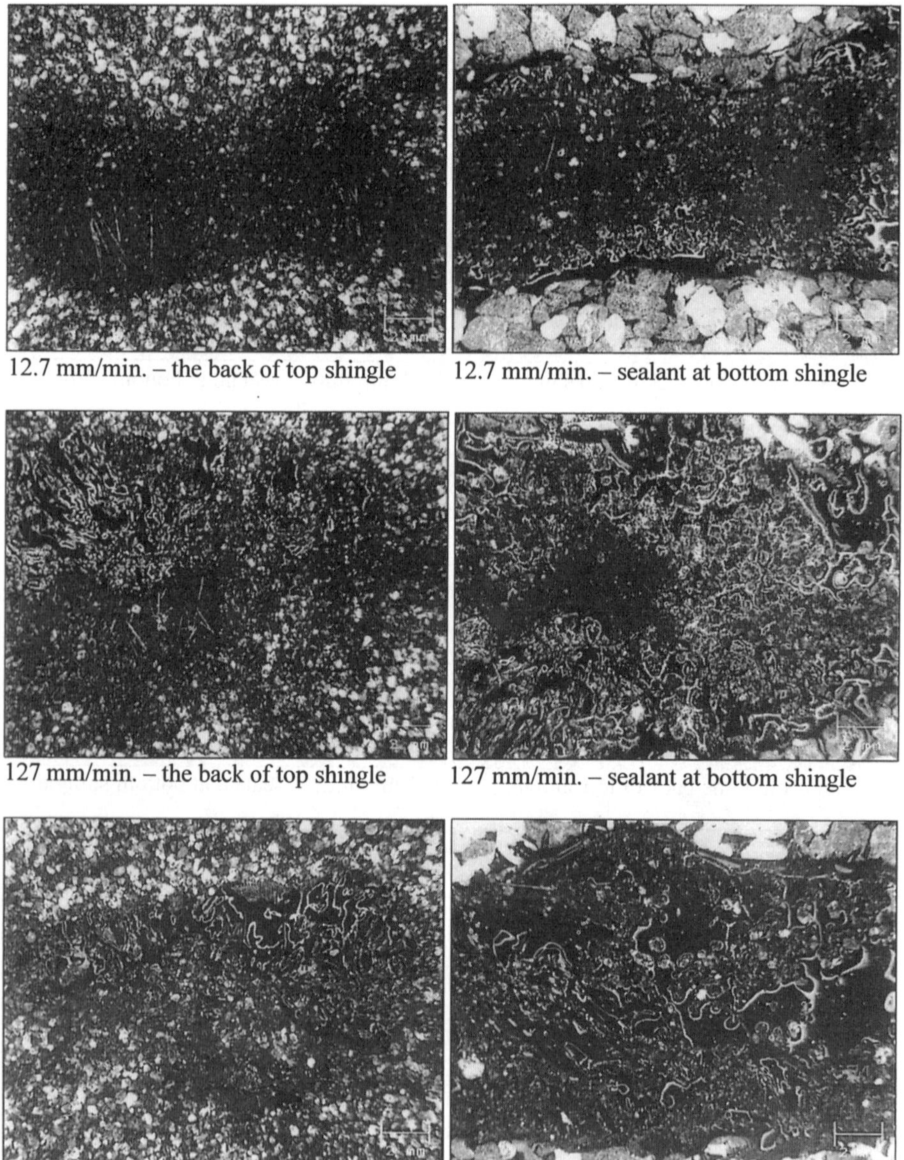

12.7 mm/min. – the back of top shingle

12.7 mm/min. – sealant at bottom shingle

127 mm/min. – the back of top shingle

127 mm/min. – sealant at bottom shingle

1270 mm/min. – the back of top shingle

1270 mm/min. – sealant at bottom shingle

FIG. 11 – *The fracture surfaces of the sealant tested by the Double-Peel Test.*

to be similar among the three different uplift tests and is also found to be independent on the shingle rigidity. This is consistent with the known cavitation phenomenon in adhesives as discovered by Gent and his coworkers [*10*], in which cavities were found to burst into existence in the bulk of the adhesive materials at a well-defined load as the load application rate increases. Although cavitation can impair the sealant strength, it is known that such bubbles can act as tiny suction cups thereby stopping the propagation of the fracture and reinforcing the adhesion energy [*10*]. As a result, the sealant strength may increase as the loading rate increases. It is important to realize that the surface must be capable of sustaining the stresses so as not to debond from the sealant or fracture internally. Hence, in the cases where uplift forces are rapidly imposed or coating asphalt becomes increasingly rigid at lowering temperatures, the sealant performance can be significantly increased due to the ability to sustain failure through cavitation.

Conclusions

The wind uplift resistance of asphalt shingles has been studied by three different mechanical test methods to examine the effects of loading configuration, shingle rigidity, and the rate of loading. A special test fixture enabling simultaneous loading on both sides of sealant has been developed to measure the shingle uplift resistance according to Perkerta's wind load model. The results show that the sealant performance is strongly affected by the various test methods and that the ASTM D6381 test can lead to significantly low measured sealant strength when compared to the results obtained from the double-peel test. As the loading rate increases, the apparent sealant strength significantly increases due to the transition of cohesive failure in asphalt coating to the adhesive failure in the sealant by cavitation. As a result, the sealant strength as measured by the ASTM D6381 may significantly underestimate the actual sealant performance when considering the case of uplift forces imposed by sudden gust winds.
The test is relatively simple to perform, even though the specimen preparation is more demanding than the ASTM D6381 peel test. The resultant sealant strength provides a good understanding for the sealant performance, particularly for the case of more rigid shingles, e.g., a multi-layer, heavy-weight shingles in cold temperatures.

The sealant performance is also found to be influenced by the rigidity of the test probe. However, due to the mismatch of the shingle exposure sizes in the samples selected for this paper, it is suggested that further study with shingle samples of similar dimensions and chemical composition be conducted to quantify such effects.

Acknowledgments

The authors would like to thank Dr. Michael J. Noone for his helpful suggestions and discussions. Also, the testing work on measuring shingle uplift resistance by Janet L. Roelofs is greatly appreciated.

References

[1] McCluer, R., "Inland High Wind Construction Issues from the Central and Eastern United States," *The Building Official and Code Administrator*, Sept./Oct. 1997, pp. 20-24.

[2] Waller, J. E., "Surviving Nature's Wrath and Human Shortsightedness," *Southern Building*, Sept./Oct. 2002, pp. 13-18.

[3] Wills, R. J., "The Impact of Hurricane Andrew on Wind Engineering," *The Building Official and Code Administrator*, July/August 1994, pp. 14-20.

[4] Minimum Design Loads for Buildings and Other Structures, *American Society of Civil Engineers*, ASCE 7-98: Revision of ASCE 7-95, ASCE, 2000, Reston, VA.

[5] Roodvoets, D. L., "Specifying Wind-Resistance in Shingles," *The Construction Specifier*, October 2002, pp. 33-38.

[6] Shaw, D. E., "ARMA's New Approach for Evaluation of Asphalt Shingle Wind Resistance," *Proceedings of the 3rd International Symposium on Roofing Technology*, National Roofing Contractors Association, Rosemont, IL, pp. 216-221.

[7] Jones, J. E. and Metz, R. E., "Comprehensive Research Program: Wind Resistance of Asphalt Shingles," *Roofing Research and Standards Development: 4th Volume, ASTM STP 1349*, T.J. Wallace and W.J. Rossiter, Jr., Eds., ASTM, 1999, pp. 112-122.

[8] Perterka, J. A. and Esterday, W. S., "Development of a Shingle Uplift Coefficient Test Standard," Final Report, CCP Project 98-1732, Asphalt Roofing Manufacturers Association, Calverton, MD, August 1999.

[9] Perterka, J. A., Cermak, J. E., Cochran, L. S., Cochran, B. C., Hosoya, N., Derickson, R. G., Harper, C., Jones, J., and Metz, B., "Wind Uplift Model for Asphalt Shingles," *ASCE Journal of Architectural Engineering*, Vol. 3, No.4, Dec. 1997, pp. 147-155.

[10] Gay, C. and Leibler, L., "On Stickiness," *Physics Today*, American Institute of Physics, Nov. 1999, pp. 48-52.

[11] McEwan, A. D., "The Peeling of a Flexible Strip Attached by a Cavitating Viscous Adhesive," *Rheol. Acta*, Vol. 5, 1966, pp. 205-211.

[12] Gent, A. N. and Tompkins, D. A., "Nucleation and Growth of Gas Bubbles in Elastomers," *J. Applied Physics*, Vol. 40, 1969, pp. 2520-2526

[13] Myers, R. R., Miller, J. C., and Zettlemoyer, A. C., "The Splitting of Thin Liquid Films. Kinematics," *J. Colloid Sci.*, Vol. 14, 1959, pp. 287-299.

[14] Kaelble, D. H., "Cavitation in Viscoelastic Media," *Trans. Soc. Rheol.* Vol. 15, No. 2, 1971, pp. 275-296.

[15] Wu, S., *Polymer Interface and Adhesion*, M. Dekker, New York, 1982, Chapter 10

A. Baskaran,[1] *M. Sexton,*[2] *W. Lei,*[2] *and S. Molleti*[3]

Pullout Resistance of Roofing Fasteners Using Different Methods

---

REFERENCE: Baskaran. A., Sexton. M., Lei. W., and S. Molleti, **"Pullout Resistance of Roofing Fasteners Using Different Methods,"** *Roofing Research and Standards Development: 5th Volume, ASTM STP 1451*, W. J. Rossiter and T. J. Wallace, Eds., ASTM International, West Conshohocken, PA, 2003.

ABSTRACT: Wind uplift rating is one of the key performance requirements for single-ply mechanically attached systems. It depends on the properties of the membrane (mechanical, physical and chemical), substrate (compressive strength and dimensional stability), and deck (thickness and deflection). An attachment system (fastener, plate and seam) integrates the above components to form an assembly. Fastener pullout resistance (FPR) from the deck is one of the essential design parameters in the system specification. As well, the FPR is an indicator of the existing deck condition in reroofing/recover applications. In the field, the FPR values are obtained using pullout testers. To quantify the accuracy of the manual pullout testers under various environmental conditions, an experimental program is in progress at the National Research Council's Dynamic Roofing Facility (DRF). From this ongoing research, this paper compares the FPR data from the field pullout tester with those obtained using the laboratory universal Instron testing machine. In the manual pullout tester, consistency of the FPR data depends on the operator. Considering this variable and to generalize the data, the above experiments are performed on two decks (steel and wood) with three different fastener types. The study finds that the field manual pullout tester underestimates the FPR data in comparison to the data obtained from the laboratory machine. This trend is found true irrespective of the selected decks and fasteners. In the roofing market, automatic pullout testers are also available that function using a battery-powered source. Selected experiments were performed to identify the influence of the pull out speed on the FPR data using an automatic pullout tester.

KEYWORDS: wind uplift, mechanically attached roofs, fastener design, roof decks, pullout testers, fastener pullout resistance

---

[1] Senior Research Officer and Group Leader, National Research Council, 1200 Montreal Road, Ottawa, ON, K1A 0R6.

[2] Technical Officers, National Research Council, 1200 Montreal Road, Ottawa, ON, K1A 0R6.

[3] Ph.D Candidate, Department of Civil Engineering, University of Ottawa, Ottawa, ON, K1N 6N5.

Introduction

Wind loading on mechanically attached single-ply roofing is a complex problem. The devastating effect of two hurricanes, Andrew (110 to 125 mph) and Iniki (90 to 100 mph), clearly directed the roofing industry in search of better engineering methods to evaluate and design roofing systems. Failure of materials from roofing systems was the single largest (95 percent) contributor to financial loss during these hurricanes [*1*]. Mechanically fastened single-ply roof systems, that appear to be the fastest growing methods for industrial roofing, face the greatest challenge of resisting the variable wind forces that act upon them [*2*]. Of particular interest is their ability to handle the effects of wind flutter and fatigue caused by uplift forces of varying intensity and direction. Figure 1 shows a typical spot-fastened mechanical single-ply system. The wind uplift forces deflect the membrane causing membrane tension and loading on the fastener. The variation of wind-induced effects with time and space creates the phenomenon of dynamic wind loading on the fastener [*3*]. Spot-fastened systems subjected to asymmetrical loading are more vulnerable to wind dynamics. When the fasteners are exposed to multi-directional wind forces, rotation and rocking occurs. These effects cause fatigue deformation on the metal deck [*4*]. Ultimately, this prolonged fatigue leads to a reduction in the pullout resistance of the fastener, a factor that must be accommodated in the design of roof systems.

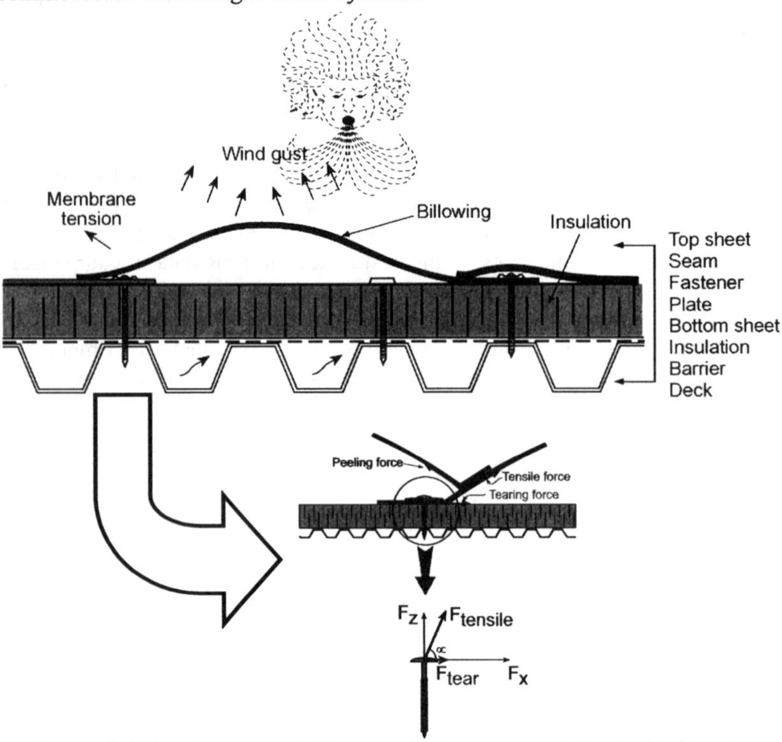

Figure 1. Wind Dynamics Effect on the Fasteners of Single Ply Roofs.

In contrast, the current industry design practices for determining fastener pullout resistance rely on tests that apply a constant static tensile force. It is understood that the development of a test procedure that can simulate the wind dynamics on fasteners is beyond the scope of the present paper. Instead, this paper attempts to determine, systematically, the accuracy of the existing field test practice. Need for such evaluations is critical due to several reasons, some of them highlighted below:

- For new construction design, typically, the FPR values are taken from the fastener manufacturer's specifications, which provide the FPR data of a fastener for various deck (steel, wood and concrete) applications. There are differences among steel roof decks [*5*], depending on their grades. For example, ASTM A 653 Standard Specification for Steel Sheet, Zinc Coated (Galvanized) or Zinc- Iron Alloy – Coated (Galvanized) by the Hot Dip Process, Grade 80 (minimum yield strength 80 ksi–550 Mpa) and Grade 33 (minimum yield strength 33 ksi–230 Mpa) are both labeled as 22 Ga steel decks and are used interchangeably across North America. FPR value of a fastener tested on decks of Grade 80 and Grade 33 can vary as much as 50%.
- Similarly, there is a significant variation in the FPR data obtained for a fastener between plywood deck and oriented strand board [*6*].
- In most of the re-roofing applications, the deck properties cannot be easily determined. In those conditions, one of the critical design parameters is the FPR data.

To quantify the accuracy of the manual pullout testers under various environmental conditions, an experimental program is in progress at the National Research Council's Dynamic Roofing Facility (DRF). From this ongoing research, this paper compares the FPR data from the field pullout tester with those obtained using the laboratory universal Instron testing machine. In the manual pullout tester, consistency of the FPR data depends on the operator. Considering this variable and to generalize the data, the above experiments are performed on two decks (steel and wood) with three different fastener types. The study finds that the field manual pullout tester underestimates the FPR data in comparison to the data obtained from the laboratory machine. This trend is found true irrespective of the selected decks and fasteners. In the roofing market, automatic pullout testers are also available that function using a battery powered source. Selected experiments were performed to identify the influence of the pull out speed on the FPR data using an automatic pullout tester.

Engineering Properties of the Selected Materials

Decks

Steel: 22 - Gauge (0.76 mm thick) galvanized corrugated sheets of 610 mm long by 915 mm (24 in. by 36 in.) wide were used as steel deck specimens. Young's modulus (Es) value of the steel is approximately equal to 200000 MPa (29 x 10^3 ksi) and had minimum yield strength 230 MPa (30 ksi). The yield strength values were not measured for the present study. These values are reported as specified by the manufacture in accordance with the ASTM A 653 specification for the metal decks.

Wood: 12 mm (1/2 in) thick plywood sheets, 1219 by 2438 mm (48 by 96 in) were used as the wood deck specimens. Young's modulus (Ew) value of the wood parallel to the grain, Ew = 10000 MPa (1451 ksi).

Fasteners

Three different fastener types were tested and their characteristics are listed in Figure 2. All installed fasteners were 130 mm (5.1 in.) long with a thread length of 102 mm (4 in.).

- Fastener #1 had a shank diameter of 4.572 mm (0.18 in.) with industry classification as #15. According to the manufacture specification, it is fabricated using hardened carbon steel with CR-10 fluorocarbon coating.
- Fastener #2 had a shank diameter of 4.37mm(0.17 in.) with industry classification as #14. It is fabricated using carbon steel with epoxy electro deposition coating.
- Fastener #3 is similar to Fastener #1 in size and is fabricated by a combination of techniques similar to that of Fastener #1 and Fastener #2, i.e., using carbon steel with CR-10 fluorocarbon coating.

As shown in Figure 2, differences exist among the fasteners in the head and tip design that can contribute to the pullout resistance. Close examination of the fastener tips also reveals the differences among the fasteners' thread design, namely distance between the thread and its angle to the shank. Nevertheless, the same fasteners were tested for both wood and steel deck, in an attempt to establish FPR comparisons between them.
All the materials were obtained as on shelf items through typical industry distribution channels.

Instron Machine (IM) - Experimental Set Up and Data

Steel Deck Samples

Samples tested were cut from 22 - Gauge (0.76 mm thick) galvanized corrugated sheets of 610 mm (24 in.) long by 915 mm (36 in.) wide. The sizes of the samples were 178 mm (7 in.) wide by 305 mm (12 in.) long. It was necessary to use these dimensions, since samples tested with shorter lengths 89 by 152 mm and 152 by 152 mm (3.5 x 6 in. and 6 x 6 in.) provided inaccurate results owing to excessive sheet deformation. On the other hand, samples tested with 305 by 305 mm (12 in. x 12 in.) do not affect the pull out resistance of the fasteners. As shown in Figure 3, the existing IRC/NRC facility can test samples with dimensions up to 457 by 610 mm (18 x 24 in.). Steel deck specimens were prepared by installing fasteners through the upper horizontal section of the deck's profile, the so-called top flange or male rib. The fasteners were installed using a drill press. Special precautions were taken during the installation process to ensure that no bending of the sheet occurred and that the pressure applied during each application remained fairly constant. In each case, the fasteners were centered with respect to length and installed with a minimum penetration depth of 19 mm (3/4 in) to a maximum depth of 25mm (1 in).

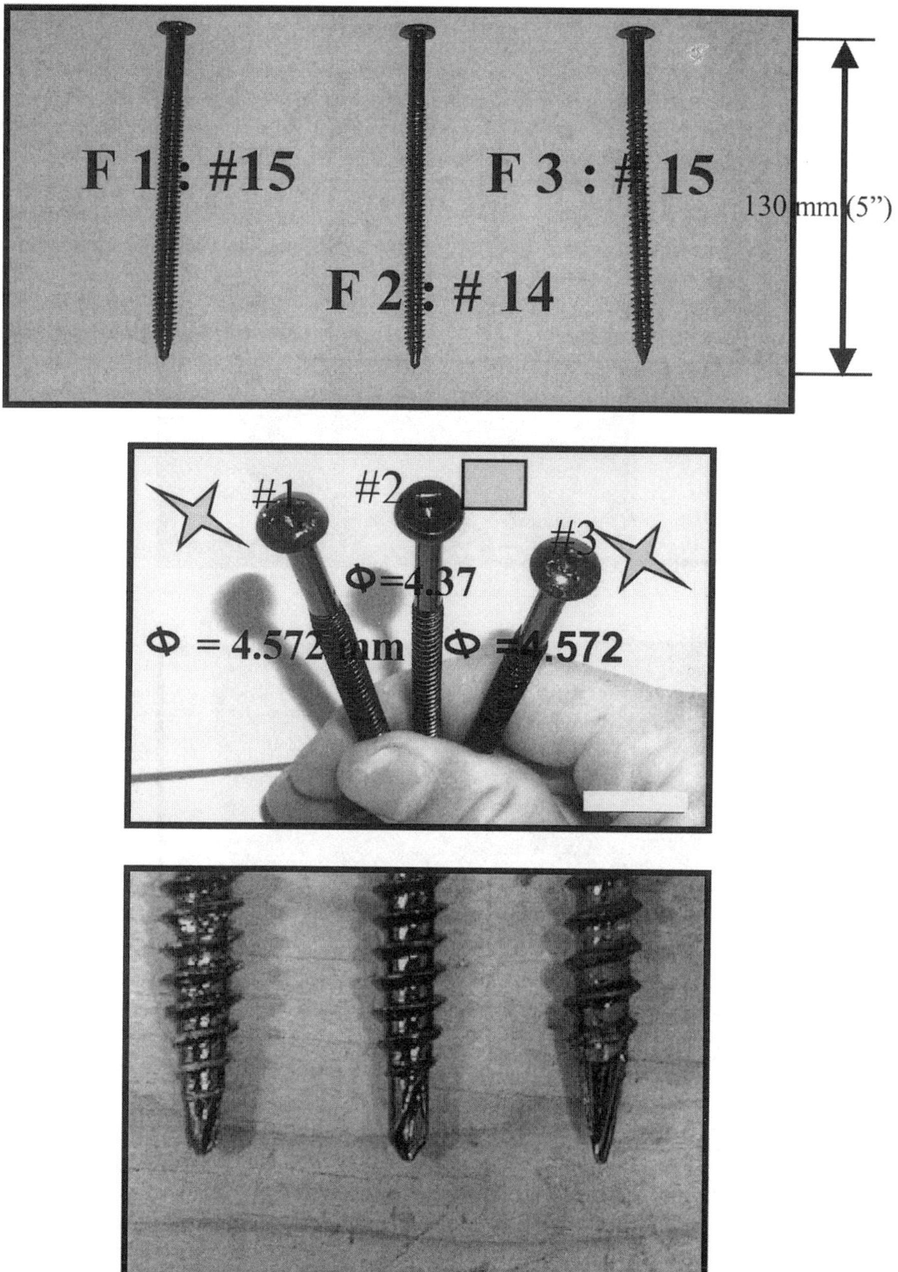

Figure 2: Characteristics of the Tested Fasteners

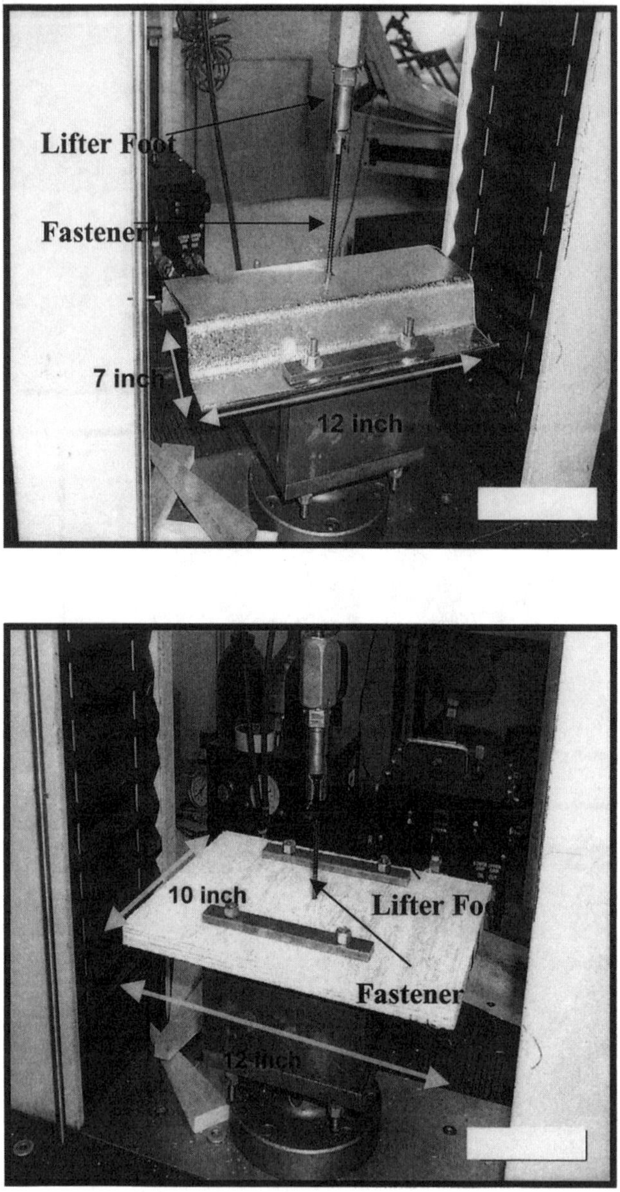

Figure 3. Experimental Set-up for Instron Machine

Wood Deck Samples

Samples of 254 by 305 mm (10 x 12 in.) in size were prepared from 1219 by 2438 mm (48 x 96 in) sheets of 12 mm (1/2 in.) thick plywood. With wood decks, fasteners were installed perpendicular to the grain at penetration depths of 38 mm (1.5 in.). Penetration depth refers to the distance from the topside of the wood to the tip of the fasteners.

For both samples, an attachment base was designed so that the samples were held down along its edges. This attachment system was connected to a computer-controlled Instron Model 4502 testing machine. A constant upward force was applied by pulling the fastener at the 3 different rates of:

- Speed 1- S1 - 40 mm/min (1.6 in./min)
- Speed 2- S2 - 48 mm/min (1.9 in./min)
- Speed 3- S3 - 64 mm/min (2.5 in./min)

While the experiment is running, the software records the load vs. displacement curve in kilograms and millimeters. At the completion of the test, data were stored in an ASCII file for further analysis.

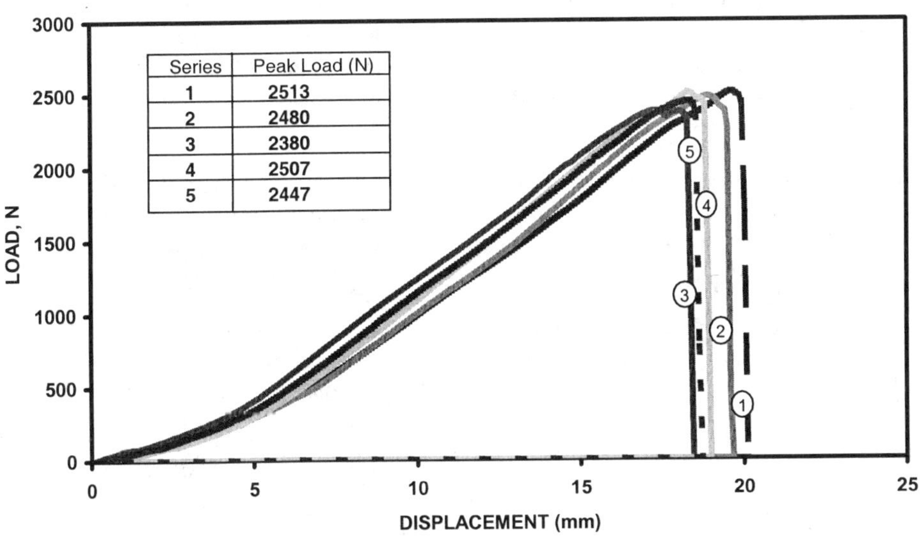

Figure 4. Typical Load - Displacement Curve from Instron Machine

In Figure 4, typical load versus displacements recorded during testing of five different steel deck samples are compared. The graph refers to results obtained from testing Fastener #1at a pulling speed of 40 mm/min (1.6 in./min). During testing, the fastener was gradually pulled out of the deck causing sheet deformation along the fastener's perimeter or area of influence. Eventually the fastener thread pulled out of the deck. In the case represented in Figure 4, the test was terminated immediately after the initial pullout of the fastener's thread from its deck engagement. Continuation of the test showed significant reduction of pullout resistance following the initial failure. The sheet deformation around the fastener was about 4 mm vertically. Slight bending and damages in terms of coating loss on the threads of the fastener were also observed. The connection failure involved both the fastener and the steel deck.

FPR data obtained from IM is shown in Figure 5 for steel and wood deck samples. Data at three different pull out speeds for three different fasteners are compared. The plotted FPR data represents the average value of more than ten independent pullout tests. The standard deviations are also calculated and shown in the figure. The following observations are noted:

- In the case of the steel deck samples, Fastener # 1 yields 2504 N (563 lbf) as FPR, the highest FPR of all the 3 fastener types tested.
- Fasteners # 1 and # 3 have the same head design and shank diameter (4.57mm) and are both classified as #15 by the industry, however, Fastener # 1 performed better than Fastener # 3. This is due to the fact that there is a difference between the tip and thread design of Fasteners # 1 and # 3. These differences contributed to the fastener's engagement with the deck, thereby affecting the FPR value.
- For the steel deck samples, using the average data, a maximum difference of 62 N (14 lbf) was calculated between the different pulling speeds. This is true for all three different fasteners.

Further scrutiny of the FPR results of wood and steel decks revealed the following:

- FPR does not vary between the testing samples in the case of steel decks. This is not true with wood decks. The maximum standard deviation for steel deck samples was 62 N (14 lbf) whereas in the case of wood deck it was 686 N (27 lbf).
- Even though steel is much thinner than wood (0.76 mm in comparison to 12 mm), fastener pullout loads are higher with steel decks. Two factors contribute to the higher pullout load with steel deck. First, the Young's modulus value of steel is high, Es = 200000 MPa (29 x 10^3 ksi), compared to that of wood, Ew (wood modulus parallel to the grain) = 10000 MPa (1451 ksi). Secondly, the steel deck gains a certain percentage of the pullout load by deflecting before the fastener fails. This was not observed with the wood deck.

Similar experiments were also carried out using OSB (a machine made product from wood flakes) deck samples instead of plywood. Overall, FPR values were greater for plywood than OSB. This is found to be true regardless of the fastener pulling speed. The difference between the two materials can be attributed to the additional strength contributed by the grain of the plywood. Plywood grains, having natural bonding, allowed the load to be more distributed thereby increasing the pullout loads. OSB, on the other hand, has low bonding between the grains. This causes a reduction in the FPR.

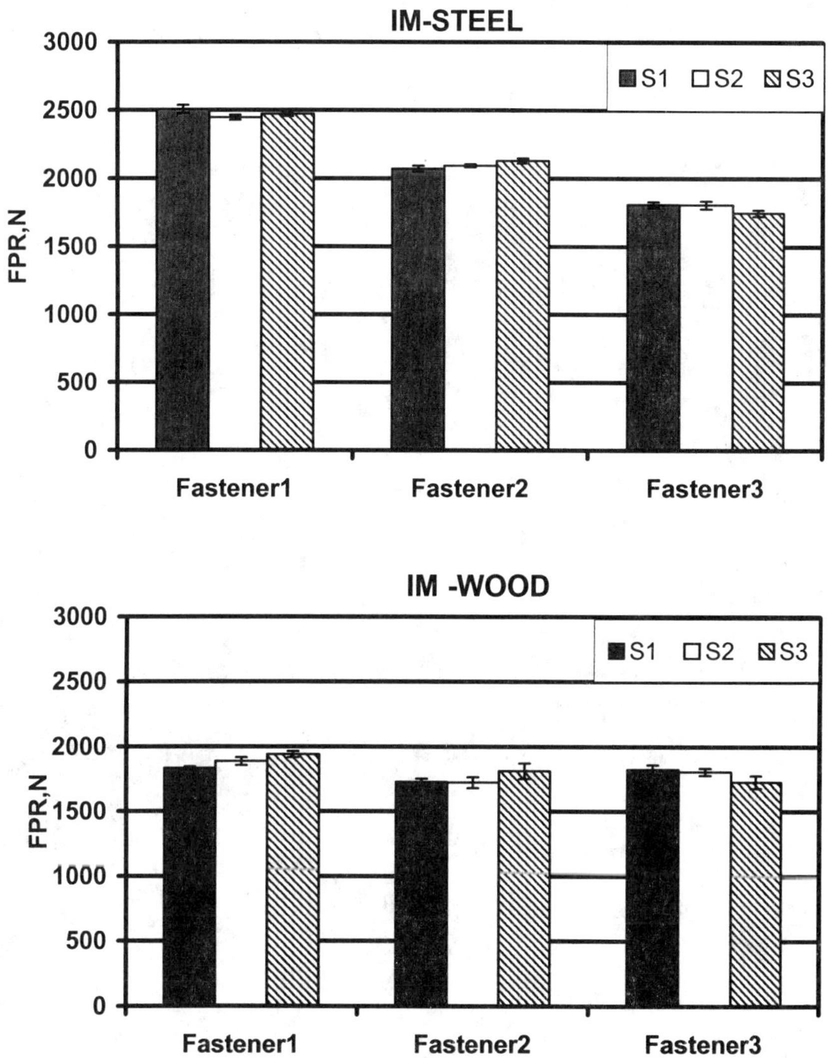

Figure 5: Pullout Resistance Data from Instron Machine (IM)

Portable Automatic Machine (PAM) – Experimental Set Up and Data

For the portable machines, an entire sheet of roofing deck was used for testing. However, the decks used were taken from the same manufactured set as those used for testing with the IM. As shown in Figure 6, both deck samples were placed on a 152 mm by 152 mm (6 x 6 in.) wood beams to provide the space needed for the fastener penetration depth and stability during the testing.

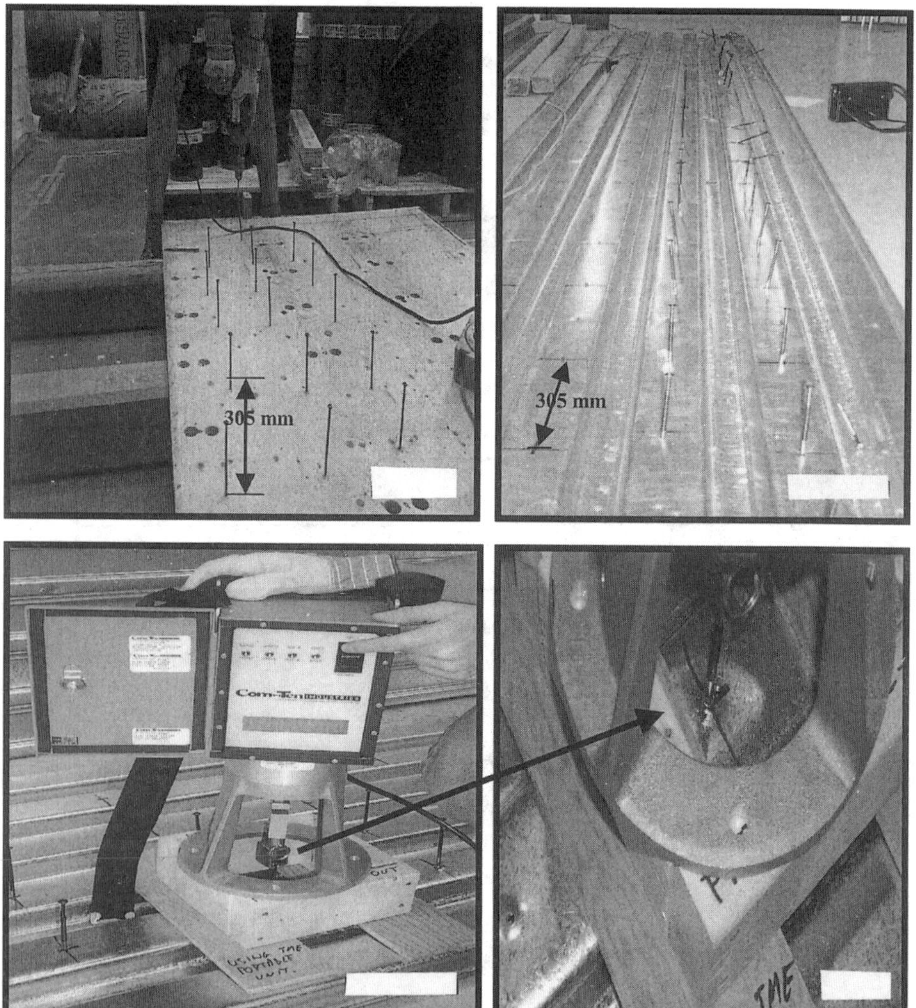

Figure 6: Experimental Set-up for Portable Automatic Machine(PAM)

Fasteners were installed on the same deck sheet so that they were far enough away from each other to avoid influencing the engagement of adjacent fasteners during testing, when the deck experienced some deformation. Therefore a distance of 305 mm (12 in.) was provided between two fasteners.

A wood base support was attached to the PAM base for two reasons:

1. First, to get enough clearance between the fastener head and PAM tester when using a 127 mm (5 in.) long fastener.
2. Second, to provide a flat surface for the PAM base to avoid any tilt of the machine during testing.

Test performed without the wood base affected the FPR values for the above-mentioned reasons. Moreover, care should be taken such that the operator is not preventing the localized deck deformation during testing. This is more critical with the steel deck samples than with the wood decks tested. Another factor that can affect the FPR data was the alignment of the fasteners. As shown in Figure 6, care should be taken when attaching the fastener head to the pullout tester, such that the fastener remains vertically aligned. FPR data significantly varies between an inclined fastener and a vertical one. This difference in values also varies depending on the fastener thread design.

PAM has three pre-set pullout speeds (40, 48 and 64 mm/min) that are similar to the speeds used in the IM. It can display the load cell data either in newton, kilogram or pound. However, it stores only the maximum pullout resistance for the test. To run the PAM, according to the manufacturer, one can use the supplied battery or an AC/DC converter. During the present study the battery was not strong enough to allow many tests continuously and consistently. After about 15 pullout tests, the machine could not maintain the required speed and the maximum pullout resistance values were not consistent. Therefore, to avoid this limitation, all experiments were conducted using an 110V-12V converter.

Figure 7 shows the FPR data on steel deck samples obtained from PAM. Differences in the fastener pulling speed does not affect the FPR values of the fastener. However, the three different fasteners each showed different values of FPR. The difference between fastener # 1 and # 3 was more than 814 N (183 lbf) (refer to Figure 5 in the case of IM, it was 698 N (157 lbf)). Discussion with the PAM manufacturer revealed that the output signal from load cell used in the PAM might have been interfering with the excitation signal. To recalibrate the load cell and to remove the signal interference, the PAM has been returned to the manufacturer several times. As the issue continued, it was decided to exclude the PAM data from the further comparison with the IM and error analysis.

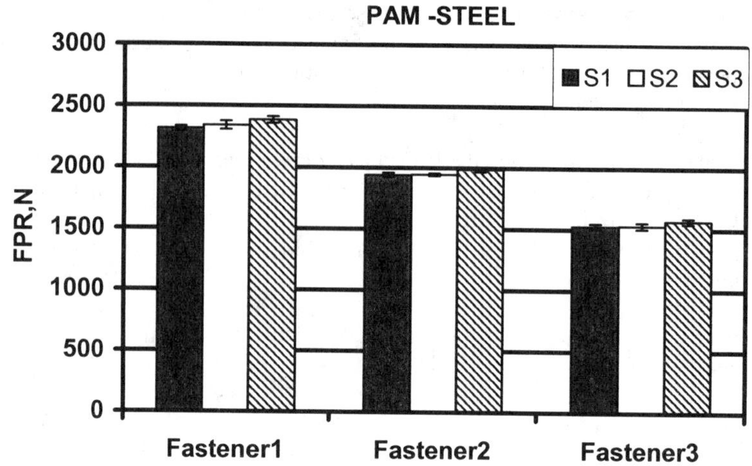

Figure 7: Pullout Resistance Data from Portable Automatic Machine (PAM)

Portable Manual Machine (PMM) - Experimental Set Up and Data

For testing with the PMM, the same samples and set up as for the PAM were used. Figure 8 shows the components of the PMM and testing in progress. Tensile forces are applied against the fastener by rotating the capstan arm. The power nut and screw arrangement transforms the rotational force into vertical tensile force. As shown in the figure, similar to PAM, a wood base attachment was used during the PMM operation. One major differences of the PMM in comparison to the other methods is the operator involvement. Three operators, two male and one female, were selected. The senior male operator had previous operating experience of the machine whereas the female and the other male operators were spectators during the pullout testing process and had a trial run before the actual testing.

Figure 9 shows the FPR data from the PMM both for steel and wood deck samples. In comparing the performance of different fasteners for the steel deck, it is clear that Fastener # 1 performs better than the others. A similar observation had also been noticed with the wood deck samples. However, the operator influence is clearly evident from the data. This influence is more pronounced in the wood deck samples then the steel deck samples. As shown, there is no consistent FPR variation amongst the different operators for the different fasteners tested. In the case of the wood deck samples, for Fastener # 1, operator 3 obtained higher FPR values than the other two operators. Similarly, operator 1 yielded the highest FPR values for Fasteners # 2 and # 3. Overall, operator dependence is more evident in the wood deck samples than the steel deck samples. This is also evident if one compares the calculated standard deviation values.

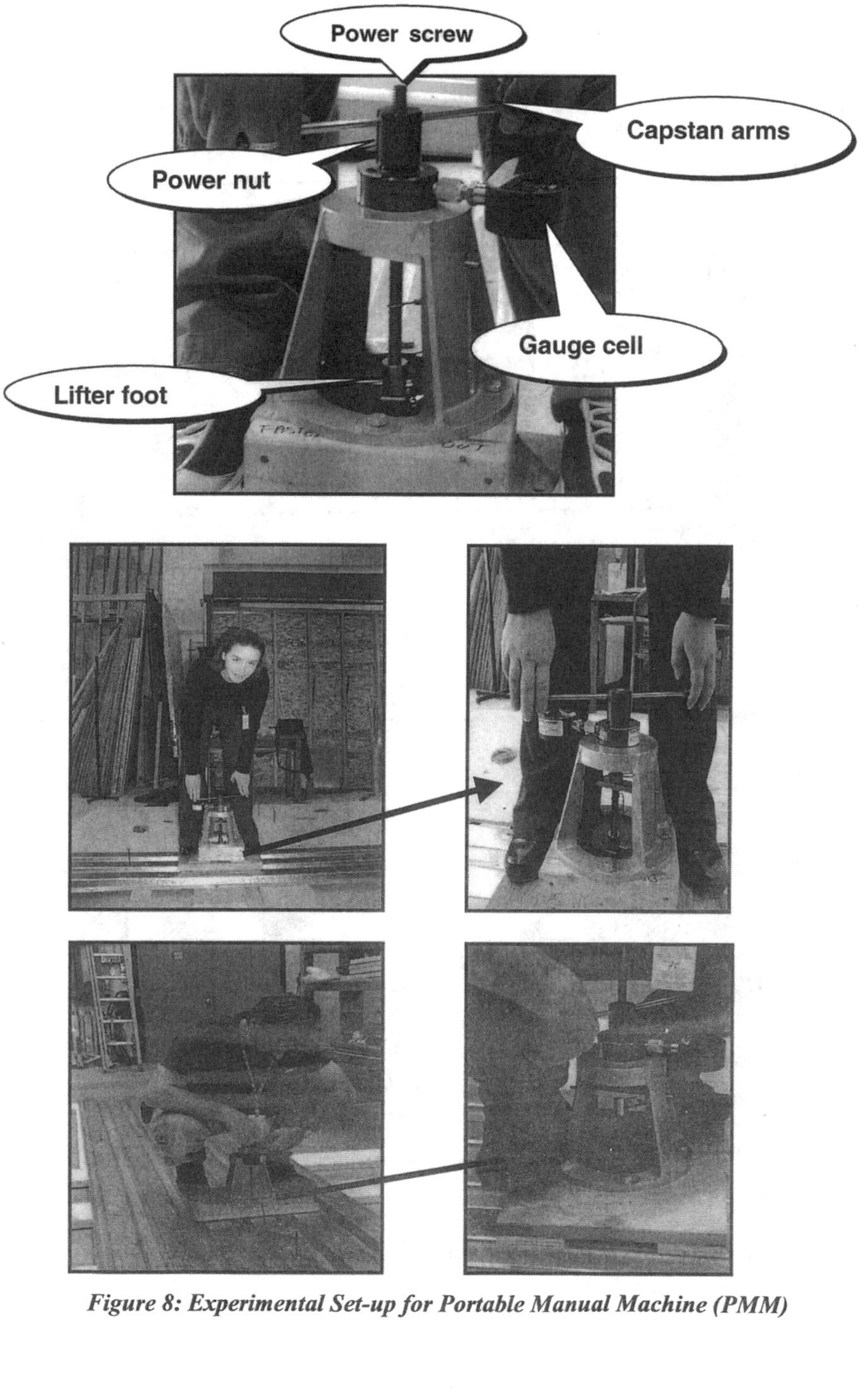

Figure 8: Experimental Set-up for Portable Manual Machine (PMM)

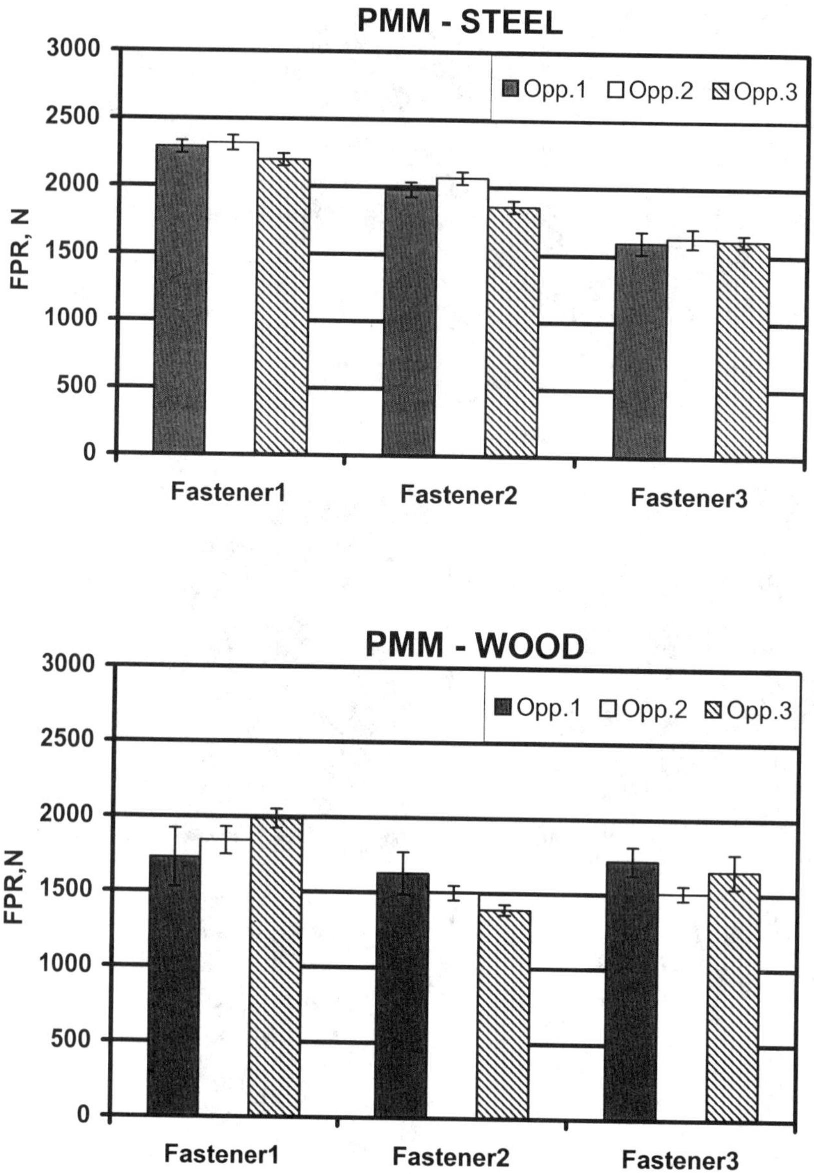

Figure 9: Pullout Resistance Data from Portable Manual Machine (PMM)

Comparison of Fastener Pullout Resistance

Figure 10 compares the FPR data obtained from the IM with that of the PMM. Data from steel and wood deck samples are presented for three different fasteners that are tested. In the case of the IM, the average data from three different speeds are compared with the average data obtained from three different operators of the PMM.

- For steel deck, a maximum deviation of about 222 N (50 lbf) between PMM and IM was noticed during the testing of Fastener # 1.
- For wood deck, a maximum deviation of about 330 N (70 lbf) between PMM and IM was noticed during the testing of Fastener # 2.
- For all fastener types, standard deviation values are higher for the steel deck samples than the wood deck samples.

To establish the accuracy of the PMM, both experimental error and deviation from IM are calculated as shown in Table 1. The experimental error represents the PMM and IM difference in the standard deviation values for the respective fastener and deck type. Deviation error represents difference in the average value between the PMM and IM. The steps involved in the calculations are shown below:

$$\Delta E = \frac{1}{3}\sum_{i=1}^{3}\left(SD_{PMM} - SD_{IM}\right)_i \quad (1)$$

$$\Delta D = \frac{1}{3}\sum_{i=1}^{3}\left(AVG_{PMM} - AVG_{IM}\right)_i \quad (2)$$

$$AVG = \sum_{j=1}^{n}\left(\frac{FPR_j}{n}\right) \quad (3)$$

where:

SD	Standard Deviation
AVG	Average of the FPR value
IM	Data from Instron machine
PMM	Data from portable manual machine
n	Number of samples tested
i	Data from three speeds (40, 48 and 64 mm/min) for the IM where as for the PMM data from three operators (two male and a female)

By algebraically adding the values of the equations (1) and (2), one can obtain the deviation of the PMM from IM. As tabulated, in the case of steel deck samples the experimental errors are less than deviation errors and this is not true for the wood deck samples. Regardless of the scenario, the PMM FPR data was always less than that of the IM. The PMM underestimates the FPR on the steel deck samples by 191–276 N (43-62 lbf) depending on the fastener type whereas in the case of the wood deck samples, the difference ranged from 231–347 N (52-78 lbf), depending on the fastener type.

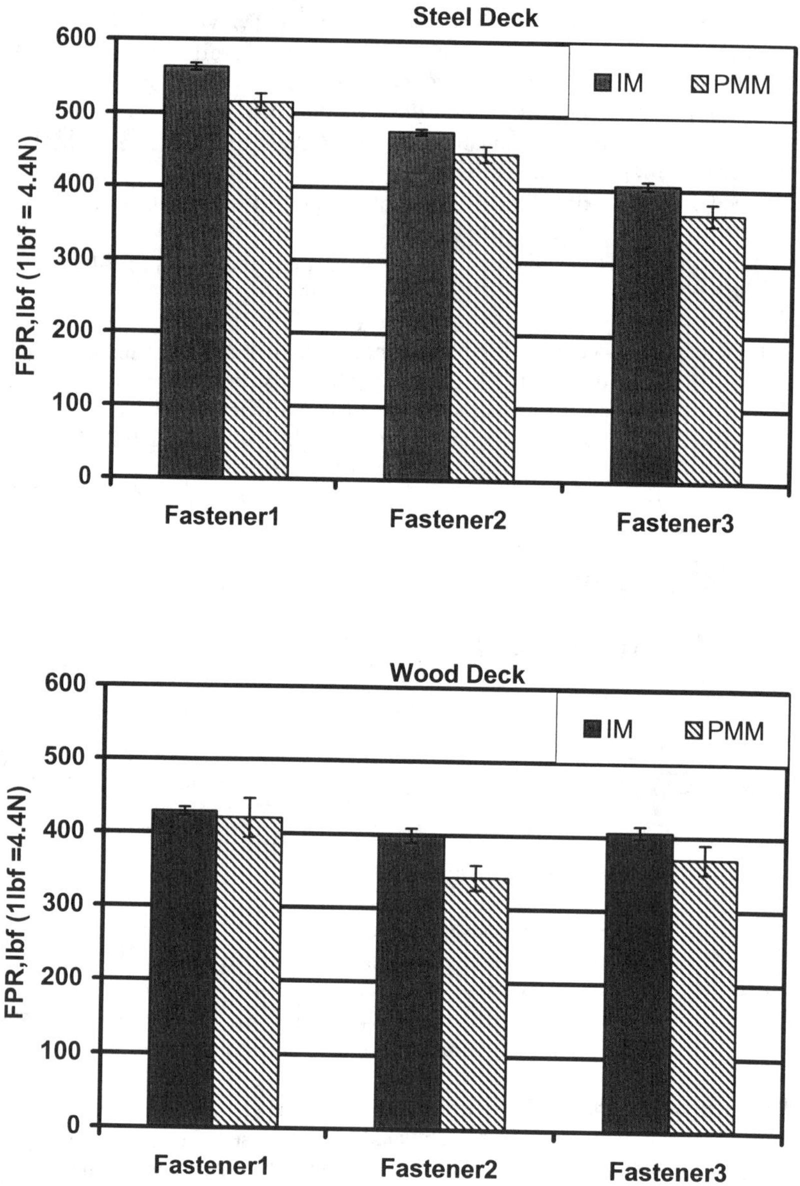

Figure 10: Pullout Resistance Comparison between IM and PMM

TABLE 1: Deviation of the PMM from IM

Fastener #	Steel Deck					
	ΔE		ΔD		Déviation	
	lbf	N	lbf	N	lbf	N
Fastener 1	*14*	*62*	*48*	*214*	*62*	*276*
Fastener 2	*13*	*58*	*30*	*133*	*43*	*191*
Fastener 3	*19*	*85*	*39*	*173*	*58*	*258*

Fastener #	Wood Deck					
	ΔE		ΔD		Déviation	
	lbf	N	lbf	N	lbf	N
Fastener 1	*44*	*196*	*8*	*36*	*52*	*231*
Fastener 2	*15*	*67*	*57*	*254*	*72*	*320*
Fastener 3	*22*	*98*	*36*	*160*	*58*	*258*

Summary

From the limited data presented in this paper, one can generalize the following conclusions.

- The speed has no influence on the maximum pullout resistance.
- Larger fastener shank diameter or larger fastener type does not necessarily yield higher fastener pull out resistance.
- Good homogeneity of the data was obtained from the steel deck samples. Pullout resistance values from the steel deck samples are greater than those from wood deck samples.
- Greater experimental error observed in the usage of PMM than the error from the IM.
- In field applications, overall, one should account for about 220-330 N (50 –70 lbf) underestimation in the fastener pullout resistance, when portable manual machines are used.

This paper presented a discussion that concentrated only on static pullout resistance of new materials. These resistance values are critical in determining the fastener design load with the respective deck composition. An equally important feature in fastener design is the prediction of its service life under dynamic conditions. For the steel deck, this phenomenon is influenced by various factors such as fatigue loading, deck corrosion, and thermal bridging. For wood deck, the factors influencing this phenomenon include moisture content, material non-homogeneity and the sample age. Efforts are in progress at IRC/NRC to develop proper experimental procedures to account for some of these effects and to gather experimental data to predict fasteners' service life. Development of such design data is vital in re-roofing situations to determine the strength of the existing fastening system. Supporting this view, using Instron Machine, two types of experimental results are reported, one on fatigue loading on steel decks and another on water immersion effects on wood decks [4]. To establish the accuracy of the PMM in situations that are common in re-roofing applications, experiments will be undertaken using the PMM.

Acknowledgment

For conducting some of the experiments reported in the paper, the authors gratefully acknowledge the assistance of Sophie Nabon and Hugues Magnier (Exchange students from the Ecole Polytechnique France).The presented research is being carried out for a consortium - Special Interest Group for Dynamic Evaluation of Roofing Systems (SIGDERS). SIGDERS was formed from a group of partners who were interested in roofing design. These partners included: *Manufacturers* - Atlas Roofing Corporation, Canadian General Tower Ltd., GAF Materials Corporation, GenFlex Roofing Systems, Firestone Building Products Co., IKO Industries Ltd., Johns Manville, Sarnafil, Soprema Canada and Stevens Roofing Systems. *Building Owners* - Canada Post Corporation, Department of National Defence, Public Works and Government Services Canada. *Industry Associations* -Canadian Roofing Contractors' Association, Canadian Sheet Steel Building Institute, Industrial Risk Insurers, National Roofing Contractors' Association and Roof Consultants Institute. *Research Agencies*-Institute for Research in Construction, Institute for Aerospace Research and Canadian Construction Material Centre.

References

1. Gerhardt, H.J. and Gerbatsch, R.W., "Wind Resistance of Mechanically Attached, Single-Ply Systems, Fastener Load, Safety Considerations and Optimal Fastener Patterns, A Decade of Change and Future Trends in Roofing", *International Symposium on Roofing Technology*, National Roofing Contractors Association, USA, (1991), pp 276-282.
2. Warshaw, S.W. and Hoher, K., "Mechanical Fastening of Single-Ply Roof Membranes into Steel Decks--An Engineering Evaluation, A Decade of Change and Future Trends in Roofing", *1985 Second International Symposium on Roofing Technology*, (1985), pp. 183-193.
3. Mahendran, M., "Fatigue Behavior of Corrugated Roofing Under Cyclic Wind Loading", Transactions of the Institution of Engineers, Australia, Civil Engineering, Vol. CE 32, No. 4, (1990 b), pp. 219 - 226.
4. Baskaran, B.A. and Dutt, O., "Performance of Roof Fasteners Under Simulated Loading Conditions", *Journal of Wind Engineering and Industrial Aerodynamics,* 72, pp. 389-400, 1997.
5. Graham, M.S., 2001, "Differences Among Steel Roof Decks", Professional Roofing, October Issue, National Roofing Contractor Association.
6. Chui,Y.H. and Craft, S., 2002, "Fastener Head Pull Through Resistance of Plywood and Oriented Strand Board", *Canadian Journal of Civil Engineering*, 29, 384-388.

M. M. Khan,[1] R. L. Alpert[1] and G. A. Smith[2]

Fire Tests for Non-Melting Plastic Rooflight Materials

Reference: Khan, M. M., Alpert, R. L., and Smith, G. A., **"Fire Tests for Non-Melting Plastic Rooflight Materials,"** *Roofing Research and Standards Development: 5th Volume, ASTM STP 1451,* W. J. Rossiter and T. J. Wallace, Eds., ASTM International, West Conshohocken, PA, 2003.

Abstract: A dimensionless "flame acceleration factor" (FAF) based on an existing concurrent flame spread model that includes consideration of fuel burnout, has been used to evaluate the flame spread on thin rooflight (or skylight) materials. The FAF is a function of ignition time, burning duration and chemical heat release rate obtained from bench-scale tests in the ASTM E-2058 Fire Propagation Apparatus (FPA). A direct correlation between the FAF values and the extent of flame spread over eighteen thin polymeric rooflight materials (1.0 – 3.3 mm thick and 1.4 – 4.0 kg/m^2 areal density) measured in large-scale tests in the ASTM E108 Test Apparatus has been presented in this paper. In accordance with the model prediction, the flame spread remains within the propane burner flame exposure region in E-108 tests when FAF < 0. Flame spread is distinctly beyond the exposure source when the FAF > 0, except for the case of some extensive melting thermoplastics. This model does not take the melting phenomenon into account.

A calculation procedure for safe separation distances for rooflights that support flame spread has been developed. These separation distances are based on chemical heat release rate, critical heat flux for ignition and smoke yield of the material as measured in the FPA, the length of rooflight that is burning at any given time and the flame height, and overall dimensions and orientation of potential rooflight targets. The calculated safe separation distance is found to be greater than one-half the total flame height.

Keywords: plastic rooflight, flame spread, fire tests, heat release, separation distance

Introduction

The potential for concurrent flame spread on a thin material overcoming the localized burnout of the fuel depends on a balance between the heat flux at which the fuel is heated to ignition, the rate of burning of the fuel and the total amount of fuel available for combustion. The concurrent flame spread model developed by Quintiere et al. [*1*] includes such consideration of the finite burning time of thin materials. Based

---

[1]FM Global Research and [2]FM Approvals, 1151 Boston-Providence Turnpike, Norwood, MA 02062.

on this model, a dimensionless flame acceleration factor (FAF), which presumably controls the flame spread process, has been derived. The FAF is evaluated from measurements, in bench-scale apparatus, of ignition time, burning duration and heat release rate and has been correlated qualitatively with heat release rate measurements associated with flame propagation over large-scale textile wall coverings [*2,3*].

In this paper, the concurrent flame spread model [*1*] that includes fuel burnout has been applied to the evaluation of flame spread on thin polymeric materials. A direct correlation between the extent of flame spread over thin polymeric rooflight (or skylight) materials measured in large-scale tests and the FAF based on the model [*1*] is presented. The large-scale and bench-scale tests are conducted in accordance with ASTM test method for fire test of roof coverings (E 108-96) and ASTM test method for measurement of synthetic polymer material flammability (E 2058-00), respectively [*4*]. Many roof surfaces may have combustible materials other than the rooflight which may contribute to growth of a rooflight fire. This study only addresses the issue of rooflight materials.

In addition, a calculation procedure for safe separation distances to limit flame spread between rooflights has been developed. The separation distances for rooflights that support flame spread are calculated using the chemical heat release rate, critical heat flux for ignition and smoke yield of the material as measured in the ASTM E 2058 Fire Propagation Apparatus. The other factors are also considered in the calculation, such as the length of rooflight that is burning at any given time and the flame height, and overall dimensions and orientation of potential rooflight targets.

Flame Acceleration Factor [*4*]

The concurrent (or, wind-aided) flame spread model of Quintiere et al [*1,8*] considers the potential of flame spread in terms of the ignition and burnout of surface elements as they are subjected to heat fluxes imposed by the flame and/or external sources. The model and its simplifying assumptions are described in Ref *1*. Based on this model, a dimensionless "flame acceleration factor" (FAF) for evaluation of thin materials is derived. The flame acceleration factor, a parameter defining the propensity of flame spread, is expressed as:

$$\text{Flame Acceleration Factor} = (k_f\dot{Q}''_{ch} - t_f/t_b)\text{-}1 \tag{1}$$

where t_f is a characteristic ignition time (s) for the pyrolysis front to move across the flame heat transfer region.; t_b is the burning duration (s); $\dot{Q}''_{ch}$ is the chemical (actual) heat release rate per unit surface area (kW/m^2), assumed to be constant over the burning duration [*1*]; and k_f is the proportionality factor for the approximate, linear relationship between dimensionless flame height (as the burnout begins) and chemical heat release rate, expressed as [*1*]:

$$\frac{(x_f - x_b)}{(x_p - x_b)} = k_f\dot{Q}''_{ch} \tag{2}$$

Here, x_f is the flame length (m), x_b is the burnout front length (m) and x_p is the pyrolysis front length (m), as depicted in Figure 1. The value of k_f is approximately 0.01 (m^2/kW) for the range of flame heights of interest, as suggested by Cleary and Quintiere [*5*]. The model [*1*] predicts that the rate of concurrent flame spread will accelerate only when $(k_f \dot{Q}''_{ch} - t_f / t_b - 1) > 0$, thereby demonstrating the significance of $\dot{Q}''_{ch}$ and t_f / t_b on flame spread. Decay to extinction is expected when FAF< 0.

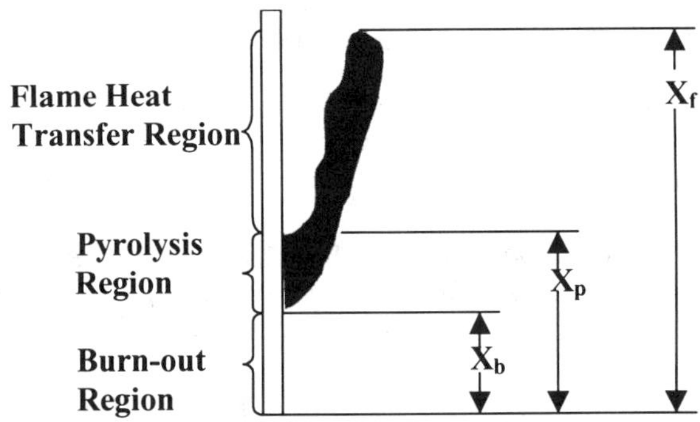

Figure 1 – *Flame spread problem* (Ref *1*).

Table 1 – *Flame Acceleration Factor and Extent of Flame Spread* [*4*].

Material	Areal Density (kg/m^2)	Thickness (mm)	$\overline{\dot{Q}''_{ch}}$ (kW/m^2)	t_f (s)	t_b (s)	$\overline{\dot{Q}''_{ch}} k_f - t_f / t_b - 1$ (-)	L_S/L_T in E-108 (-)
FRP1	2.78	1.75	142	21	111	0.23	0.54
FRP2	2.40	1.40	114	63	86	-0.59	0.308
FRP3	2.55	1.65	118	47	140	-0.16	0.308
FRP4	2.13	1.65	338	21	90	2.15	1.0
FRP5	1.37	1.0	145	17	55	0.15	0.54
FRP6	1.96	1.3	372	29	81	2.36	1.0
FRP7	1.85	1.3	416	27	69	2.77	1.0
FRP8	2.04	1.3	199	24	86	0.72	0.69
FRP9	1.93	1.3	368	28	82	2.34	1.0
FRP10	2.45	1.4	179	28	97	0.50	0.54
FRP11	3.84	2.5	362	44	142	2.31	1.0
FRP12	3.98	2.5	212	36	154	0.89	0.77
PC1	1.78	9.9	207	69	165	0.65	0.308
PC2	1.17	6.1	96	47	195	-0.28	0.308
PC3	3.30	3.25	208	85	240	0.73	0.308
PMMA	3.94	3.25	660	30	143	5.40	1.0
PET	3.60	3.25	210	43	159	0.83	0.308
PVC	1.86	1.2	45	22	224	-0.65	0.308

L_S = Flame Spread Length (m); L_T = Total Length of the Sample (3.96 m); k_f = 0.01 m^2/kW

Sample Materials

The thin polymeric materials examined in this study are glass fiber-reinforced plastic (FRP), polycarbonate (PC), polymethylmethacrylate (PMMA), polyethylene terephthalate (PET) and polyvinyl chloride (PVC), which are commonly used as rooflights (or skylights). The thickness (m) and the areal density (kg/m^2) of the materials are given in Table 1. All materials are solid sheets, except for PC1 and PC2, which are triple- and double-layered rectangular structures with the wall thickness of each layer less than 1 mm. FRP is a thermoset (char-forming material) while PC, PMMA, PET and PVC are thermoplastics.

Test Procedure and Results

Bench-Scale Tests [4]

Heat Release Measurement - Tests have been conducted in the ASTM E 2058 Fire Propagation Apparatus using a 90 mm by 90 mm sample surface uniformly coated with about one gram of carbon black powder to maximize the absorptivity of the sample surface to the incident flux. To further reduce heat losses beyond what is required in ASTM E 2058, two layers of 3.2 mm thick ceramic paper are placed on the back of the sample and then, the sides and back are covered with two layers of 0.051 mm thick aluminum foil. The insulated sample is placed on a wire mesh platform, supported on a load cell. A small pilot flame (about 10 mm long) is positioned about 10 mm above the center of the sample surface. The test is performed with co-flowing normal air (3.3 x10^{-3} m^3/s) within a quartz tube surrounding the sample in the Fire Propagation Apparatus. The sample is exposed to 50 kW/m^2 radiant heat flux and the time to ignition, t_f, and burnout or flame out time, t_{bo}, are recorded. Measurement of chemical (actual) heat release rate per unit sample surface area, $\dot{Q}''_{ch}$ (kW/m^2) as a function of time is also made, as shown in Figure 2 for the FRP6 sample, as an example. The total burning duration, t_b (s), is the difference between t_{bo} and t_f.

For thin materials, as evaluated in this study, $\dot{Q}''_{ch}$ is not constant over the total burning duration, t_b. In fact, after the onset of ignition, $\dot{Q}''_{ch}$ is highly transient in most cases, with distinct peaks only of several seconds duration. To address this problem, the average chemical heat release rate per unit area, $\bar{\dot{Q}}''_{ch}$ (kW/m^2) for the material during the active burning period (t_b) is used in the data analysis, as suggested in Ref 5:

$$\bar{\dot{Q}}''_{ch} = \frac{\int_{t_f}^{t_{bo}} \dot{Q}''_{ch} dt}{t_b} \tag{3}$$

The parameters, t_f, t_b and $\bar{\dot{Q}}''_{ch}$, along with the flame acceleration factor and extent of flame spread in ASTM E-108 tests are presented in Table 1 for all the materials tested in this study.

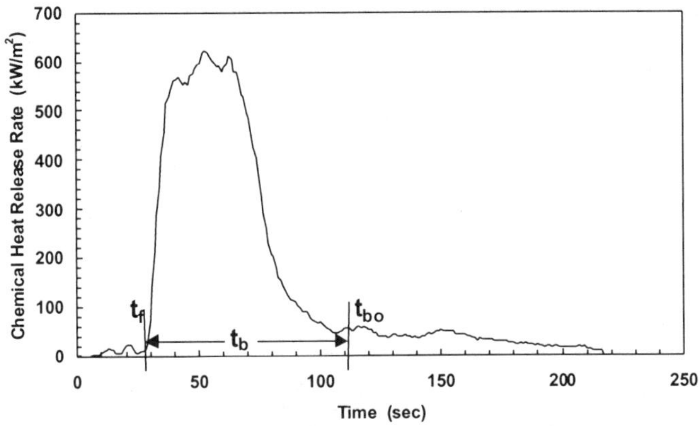

Figure 2 – *Chemical Heat Release Rate as a Function of Time for FRP6 Sample.*

Critical Heat Flux Measurement – Test conditions and sample preparation is the same as above, except the heat release rate measurement. The samples are only exposed to several values of radiant heat fluxes, and the time to piloted ignition is measured. The critical heat flux (CHF) for ignition is determined by plotting the inverse time to ignition versus the lower radiant heat flux values, where the ignition times are very large, and noting the intercept of the linear fit with the axis corresponding to infinite ignition time.

Smoke Yield Measurement – Test conditions and sample preparation is the same as the heat release rate measurement. The measurement conditions include use of standard three-color technique [*6*] with extinction coefficient obtained from the red wavelength (632.8 nm). The mass flow rate of smoke is calculated using the technique used in Ref *6*. The smoke yield is the average resulting from the ratio of the time integration (t_f to t_{bo}) of the mass flow rate of smoke (g/s) in the measuring duct to the sample mass loss (g) from the load cell measurement during the same interval, t_f to t_{bo}.

Large-Scale Tests [*4*]

Large-scale flame spread tests have been conducted in ASTM E 108 Test Apparatus with the sample holding frame work shown in Figure 3. The apparatus is designed to test roofing materials in the presence of wind-aided flames.

The framework of overall dimensions 0.91 by 3.96 m, has 38 mm x 38 mm steel angle cross-supports spaced every 0.61 m, simulating skylight frames. The frame holding the sample is tilted 23-degrees from the horizontal plane. This specific 23-degree angle, equal to the maximum allowed for the apparatus, is used to ensure that there is good flame contact with the surface of the sample. It has been shown [*7*] that for a burning surface at an angle greater than 17 degrees from the horizontal, flames attach to the surface, resulting in a concurrent or wind-aided type of burning [*7*].

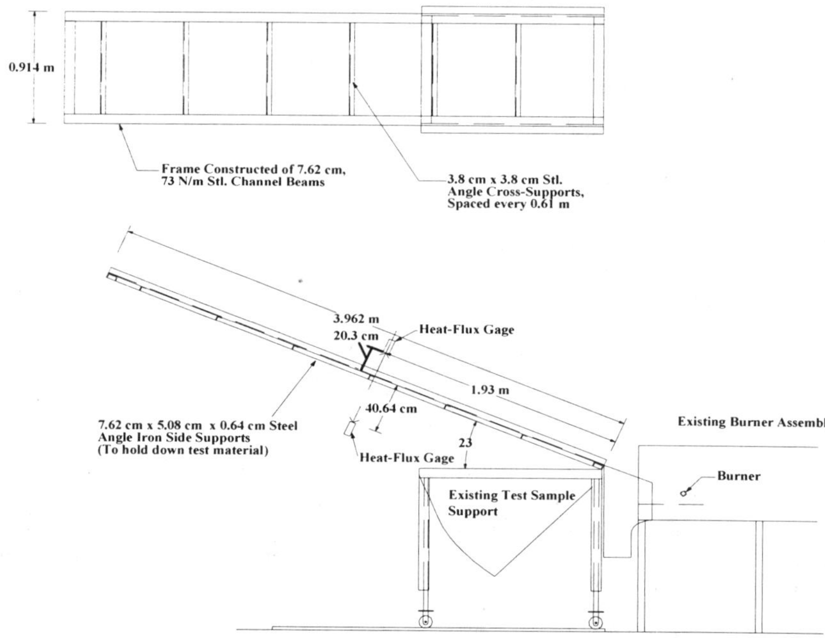

Figure 3 – *ASTM E-108 Test Apparatus as Used for Rooflight Materials.*

A forced air velocity of 5.3 ± 0.2 m/s from a fan drives a propane flame exposure of 369-387 kW over about a 1.2 m length at the lower end of the sample surface, which is 30.8% of the total length of the sample. The peak flame heat flux (as measured by a water-cooled Gardon heat flux gage mounted in an inert board) over the initial 0.61 m length of the framework is found to be between 20 and 60 kW/m^2, as shown in Figure 4. The average peak heat flux is about 30 to 40 kW/m^2, which is well above the minimum heat flux for ignition (7-20 kW/m^2) of rooflight materials tested in this study. During tests with rooflight materials, two heat flux gages, one 0.2 m above the material surface and one 0.41 m below, are placed 1.9-m from the lower edge of the sample material, to estimate flame arrival while still being beyond the influence of the exposure fire. For a Class A rating in the E 108 test, flaming is not allowed spreading beyond a 1.83-m distance from the lower edge of the sample. The tests are video taped and the propane flame source is run for 10 minutes.

The extent of flame spread in terms of the ratio of the flame-spread length (based on visual observation as well as video recording) to the total sample length is included in Table 1 for all the materials tested in this study.

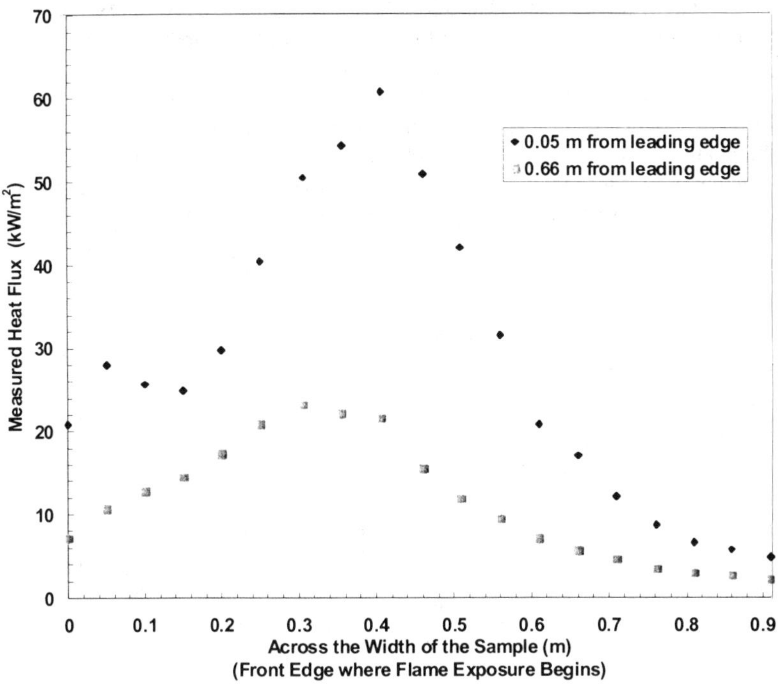

Figure 4 – *Measured Heat Flux (from the propane burner) over the Initial Length of the Sample Area in ASTM E-108 Test Apparatus.*

Discussion

Flame Acceleration Factor [4]

The flame acceleration factor (FAF), $(k_f\,\overline{\dot{Q}''_{ch}} - t_f / t_b - 1)$, and the extent of flame spread in large-scale E 108 tests (from Table 1) are plotted in Figure 5. The value of the FAF depends on the heat release rate history ($\overline{\dot{Q}''_{ch}}$) during the active burning period (t_b), the ratio of ignition time to active burning period, t_f / t_b, and the externally imposed heat flux in the apparatus where these quantities are measured. In some cases, small variations in any of these components of the FAF can have a significant effect on the prediction of flame spread. The heat flux (30-40 kW/m^2) from the exposing flame in E108 tests is somewhat lower than the external radiant heat flux (50 kW/m^2) in the bench-scale tests. Nevertheless, the prediction of flame spread beyond the 30.8% of total sample length for values of the FAF greater than zero appears to be successful, as shown in Figure 5.

It can be seen in Table 1 and Figure 5 that for the materials with negative FAF values, the flame spread remains within the exposure flame in the E 108 tests. Two FRP materials (FRP1 and FRP5), that exhibit marginal flame spread (54%-77% of the total sample length) have a FAF greater than zero but less than one. Acceleratory flame spread over the entire surface is observed for samples with FAF greater than two. These results appear to be consistent with the model prediction [*1,8*].

It should be noted that some softening or melting type of samples, such as PC1, PC3 and PET do not support flame spread beyond the exposure flame region in the large-scale tests, although the corresponding flame acceleration factors are greater than zero (0.65-0.83). The main reason for this is the melting of these samples, leading to sagging of the samples away from the exposure flame. In the bench-scale tests, the same

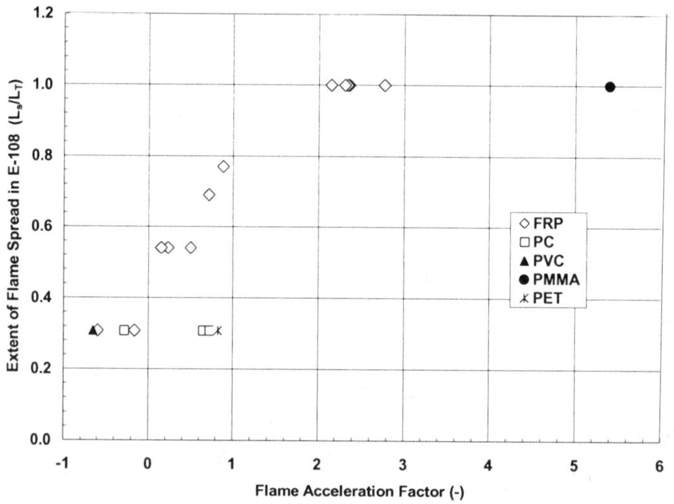

Figure 5 – *Relationship between Flame Acceleration Factor and the Extent of Flame Spread in ASTM E-108 tests* [*4*].

materials are contained in the sample holder in spite of the melting process. Of course, the flame spread model [*1,8*] was never intended to include the effects of melting thermoplastics. Note that the melting thermoplastics in Table 1 (PC2 and PVC) with FAF values of less than zero (or negative) also do not support flame spread beyond the vicinity of the exposure flame in the large-scale tests, because of their inherently low heat release rates.

Summarizing the results, it appears that the following conditions exist for FAF [4], for the thin plastic rooflight materials: when FAF < 0 or negative, flame spread remains within the exposure flame region; when 0 < FAF 1, flame partially propagates along the length of the sample, except for some melting thermoplastics; 1 < FAF 2 is an uncertain region as there are no data points; and, when FAF > 2, flame spreads rapidly over the entire sample length.

Calculation of Safe Separation Distances for Fire Propagating Rooflights

Once fire is initiated in a rooflight that supports unlimited fire propagation (FAF >2), ignition of other rooflights not contiguous with the first is assumed to be due to radiant heat flux from flames. Burning rooflight brands propagated by the wind could also cause such a remote ignition but the location of the burning material just a short distance above the roof elevation means that brands would be very limited in size, due to the low gas velocity near the base of a fire.

The radiant heat flux at a target rooflight is equal to the emissive power of the rooflight flames, multiplied by a view factor (or a configuration factor), F, which expresses how much of the source fire is "seen" by the target compared to the total view from the target. The radiant flux to a rooflight target, $\dot{q}_t''$, is thus given by:

$$\dot{q}_t'' = \dot{q}_s'' F \tag{4}$$

where $\dot{q}_s''$ is the radiant flux at the flame surface, which is governed by the path length, L_R for radiation through the flame and by the flame extinction coefficient, k_s. For acrylic rooflights (which are mainly Polymethymethacrylate – PMMA), information published by Markstein [9] indicates that the flame extinction coefficient is from 1.3 to 1.5 m^{-1}. Once k_s is known, the radiant flux from the flame surface can be expressed as:

$$\dot{q}_s'' = \dot{q}_{\max}'' \left(1 - \exp\left[-k_s L_R\right]\right) \tag{5}$$

where $\dot{q}_{\max}''$ is the radiant flux from an optically thick flame.

The extinction coefficient for materials other than PMMA can be obtained from the smoke yield, y_s, a quantity measured routinely using a procedure in the NFPA 287 version [*6*] of the FPA. Markstein [*10*] measured the extinction coefficient in the flames of pool fires for three different polymers. These measurements are correlated with the smoke yield from FPA tests of generic polymers in Figure 6. It can be seen that an estimate for k_s is given by the following:

$$k_s = 18.86 y_s + 0.92 \left[m^{-1}\right] \tag{6}$$

In the specific case of the acrylic tested in the current study, the smoke yield was measured as 0.0245. From Eq. 6, this results in an extinction coefficient of 1.38 m^{-1}.

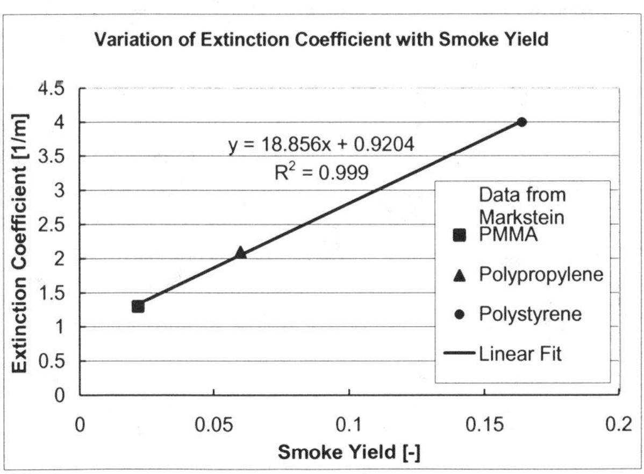

Figure 6 – *Relationship between Smoke Extinction Coefficient and Smoke Yield.*

The value of $\dot{q}''_{max}$ in Eq. 5 can be obtained from measurements made during an ASTM E 108 test of the acrylic rooflight. For this test, a heat flux gauge was mounted 0.406-m below the centerline of the specimen, at a distance of 1.93-m from the burner-exposed (lower) end of the specimen (a similar gauge 0.2-m above the specimen malfunctioned during the test). The gauge output is shown in Figure 7.

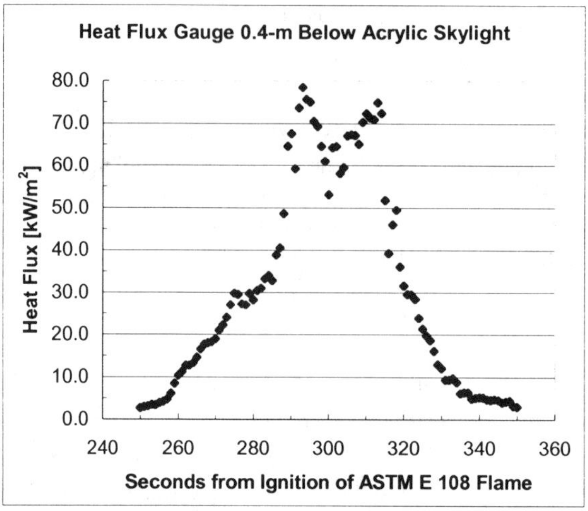

Figure 7 – *Flame Heat Flux Measurement during ASTM E 108 Test.*

From about 290 to about 320 seconds after the start of the test, the acrylic rooflight was engulfed in flames that completely filled the space between the specimen and both heat flux gauges. Hence, the path length for radiation through the flame is at least 0.406-m and possibly as much as 0.6-m if the burning PMMA is not blocking any radiation from above the specimen. Assume a total path length of 0.5-m for thermal radiation and ignore any convective heat flux. Since a heat flux of about 70 kW/m^2 is measured when the volume above the gauge is filled with flame, Eq. 5 shows that:

$\dot{q}''_{max}\left(1-\exp[-1.38\times 0.5]\right)=70$, or $\dot{q}''_{max}$ is about 140 kW/m^2 for the PMMA.

Assume that when there is a rooflight fire, the depth of the flame, L_R (path length for radiation) in the direction toward an adjacent rooflight is about one-half the rooflight width, W_f. This is the average of the flame depths at the base of the fire, $L_R = W_f$, and at the flame tip, $L_R = 0$. Note, it is assumed that rooflights are aligned with long axes parallel and that fire propagation, involving the entire rooflight width, W_f, is in the direction of the long axis, as shown in Figure 8. Hence, from these assumptions and Eqs. 5 and 6:

$$\dot{q}''_s = \dot{q}''_{max}\left(1-\exp[-(9.43y_s+0.46)W_f]\right) \quad (7)$$

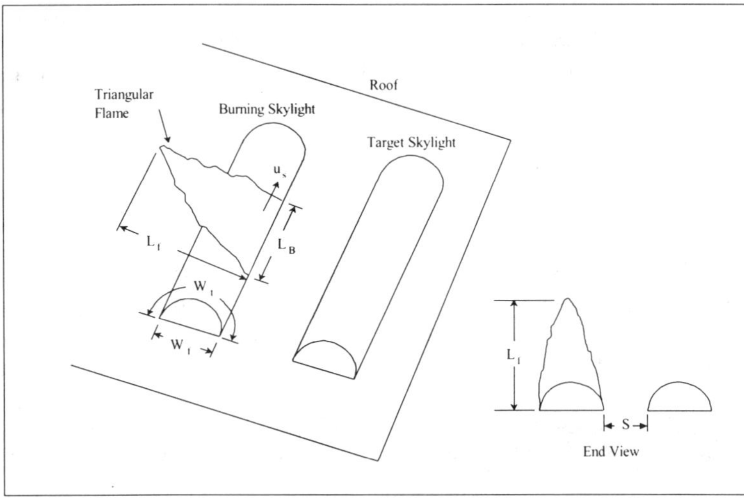

Figure 8 – *Schematic to Illustrate Radiant Exposure of Target Skylight by Moving Flames.*

The first step in computing the view factor requires the shape of the radiant fire source (mainly the fire plume) to be determined. The most critical dimensions in this regard are the length of rooflight that is burning at any given time and the flame height. The former quantity is the product of the average flame propagation velocity, u_s, and the duration of burning of a given rooflight element, t_b. An upper limit on u_s can be obtained from the rate of change of flame length observed in the ASTM E 108 test of a rooflight. Because flame propagation is wind-aided along the rooflight length axis in this type of test but the most hazardous wind direction during a fire is normal to this axis, toward an adjacent rooflight, values of u_s during a fire are likely to be much less.

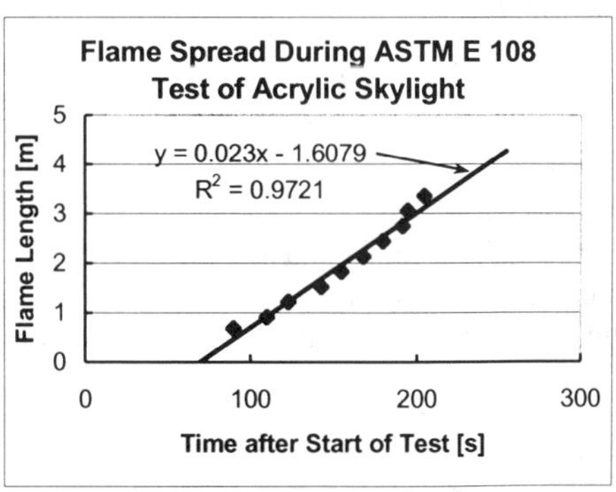

Figure 9 – *Flame Length Time History during ASTM E 108 Test of PMMA Rooflight.*

Figure 9 contains flame lengths derived from a video recording of the ASTM E 108 test of the PMMA rooflight, indicating an average flame spread velocity of 0.023 m/s during most of the propagation time period. The duration of burning, t_b, of a given rooflight element is the ratio of the heat energy content per unit area to the average heat release rate per unit area, or

$$t_b = \frac{m_f'' \Delta H_c}{\dot{Q}_{ch}''} \quad [s] \tag{8}$$

where m_f'' is the areal density [g/m^2] of the rooflight material. Assume that the average heat release rate equals the value determined from the ASTM E 2058 FPA test. Equation 8 then yields 150 s for the acrylic, in agreement with the value of 180 seconds obtained from the time between half-involvement and near consumption of the rooflight panel in the video recording.

From the product of flame spread velocity and burning duration, the maximum length, L_B, of rooflight expected to be burning at any given time is given by:

$$L_B = u_s \frac{m_f'' \Delta H_c}{\dot{Q}_{ch}''} \tag{9}$$

For the PMMA, Eq. 9 yields a burning length value of 0.023 m/s times 150 s, or about 3.5-m, which will be the source length to be used for calculation of the radiation view factor.

A flat, vertical area thermal radiation source is illustrated in the Figure 8 schematic as the result of projecting the rooflight flame onto a plane. The view factor, F, from such a source to a target at a separation distance, S, depends on the orientation and elevation of the target. The maximum view factor results from a target that is vertical (i.e., parallel to the emitting plane) and at an elevation at the mid-height, opposite the centerline of the source. Although a rooflight target may not always have such a position (especially for horizontal rooflights flush with the roof surface), the worst case position will be assumed. To obtain the following explicit formula for the view factor in terms of the separation distance, it is assumed that the radiation source is a flat disk, with the target on the disk axis:

$$F = \frac{A_s}{A_s + \pi S^2} \tag{10}$$

where A_s is the projected surface area of the rooflight flame. For burning rectangular rooflights, the fire plume tends to form a conical shape, which would appear from the target as a vertical plane-triangle. The surface area of such a triangular source is one-half the flame height, L_f, times the length of the base, which is the burning length, L_B. Hence, Eq. 10 becomes:

$$F = \frac{L_f L_B}{L_f L_B + 2\pi S^2} \;;\text{ or }\quad S = \sqrt{\frac{L_f L_B (1-F)}{2\pi F}} \tag{11}$$

The magnitude of the flame height, L_f, for a burning rectangular rooflight of length, L_B and plan width, W_f, can be obtained from the well-known Heskestad [11] correlation if the effective fire diameter is taken as the plan hydraulic diameter of the burning rooflight:

$$L_f = 0.235\left(\dot{Q}''_{ch} W_t L_B\right)^{2/5} - 2.04\frac{W_f L_B}{W_f + L_B} \tag{12}$$

where $\dot{Q}''_{ch}$ is the chemical heat release rate per unit burning surface area of rooflight and W_t is defined as the total dimension of the exposed rooflight surface in the direction perpendicular to the exposed rooflight length. The exposed rooflight length is assumed to be identical to the plan length, L_B. For example, if the rooflight is triangular in cross sectional shape, W_t is the sum of the lengths of the two sloping sides of the triangle exposed on the roof. If these sloping sides are equal in length and the height of the triangular shape above the rooflight base is H, then $W_t / W_f = \sqrt{(2H/W_f)^2 + 1}$. On the other hand, if the rooflight crosssection is a semi-circle (i.e., height one-half the plan width), then $W_t/W_f = \pi/2$. It is important to note that Eq. 12 is applicable to any shape of cross section, whether triangular, circular, etc., as long as W_t/W_f is known.

The magnitude of $\dot{Q}''_{ch}$ in Eq. 12 will be taken as the average chemical heat release rate per unit specimen area exposed to a 50 kW/m^2 radiant flux during the ASTM E 2058 FPA test. This exposure flux should be fairly representative of the heat flux in real-scale rooflight fires.

To prevent ignition of a target rooflight, the heat flux at the target, $\dot{q}''_t$, must be such that ignition will not occur during the duration of burning of the exposing rooflight. Ignition data obtained for the acrylic in the ASTM E 2058 apparatus shows that ignition time, t_f is given by the following:

$$\sqrt{\frac{1}{t_f}} = \frac{\dot{q}''_t}{279} + 0.007 \tag{13}$$

From Eq. 13, the ignition time will be at least 240 s (4 minutes, compared to a burning duration of 3 minutes) if the heat flux is no greater than 16 kW/m^2, about twice the critical heat flux, *CHF*, for this rooflight material. Generalizing this result to other rooflight materials, assume that the rooflight will not ignite during the duration of burning, t_b, if the target flux is less than or equal to 2 times the *CHF*. A more precise critical value can be obtained for individual rooflight specimens from ignition data similar to that in Eq. 13.

Using Eqs. 4 and 7, with a required target flux of 2 x *CHF*, the maximum view factor that prevents remote rooflight ignition is given by:

$$F = \frac{\dot{q}''_t}{\dot{q}''_s} = \frac{2CHF}{\dot{q}''_{max}\left\{1 - \exp\left[-\left(9.43 y_s + 0.46\right)W_f\right]\right\}} \tag{14}$$

The safe separation distance, *S*, now can be obtained from Eq. 11 in terms of L_B given by Eq. 9, L_f given by Eq.12, and F given by Eq. 14.

The safe separation distance between adjacent rooflights is given by Eq. 11 in the absence of wind effects. However, calculation of view factors for flames tilted at various angles by the wind has shown that the effect of the wind in increasing the view factor at the target will be minimal if the effective height of the vertical radiating source, which is one-half the flame height, L_f, is less than the separation distance between the source and the target. This can be explained by the fact that when the separation distance is greater than the source height, the main effect of the wind in tilting a flame is to reduce the apparent height of the source, with reductions in distance between source and target being negligible.

Examples of the results of safe separation distance calculations for the acrylic and an FRP rooflight that supports fire propagation are given in Table 2, based on the assumption that the radiant flux from optically thick flames is 140 kW/m^2 for the second rooflight as well. The rooflight cross-section is either flush with the roof ($H = 0$) or a symmetric triangle of height, H, above the roof. Other cross-sections, such as a cylindrical dome, could be handled just as easily.

Table 2 – *Examples of Separation Distance Calculations for PMMA and FRP4.*

Material	Fuel Plan-Width, W_f	Center Height above Base, H	$\dot{Q}''_{ch}$	Flame Velocity, u_s	Areal Density	ΔH_c	Smoke Yield y_s	Critical Heat Flux, CHF	Ratio, Separation Distance to Plan Width, S/W_f
	[m]	[m]	[kW/m^2]	[m/s]	[g/m^2]	[kJ/g]	[g/g]	[kW/m^2]	[-]
PMMA	1	0	660	0.023	3940	23	0.0245	8	2.41
PMMA	2	0	660	0.023	3940	23	0.0245	8	1.69
PMMA	4	0	660	0.023	3940	23	0.0245	8	1.07
PMMA	1	0.5	660	0.023	3940	23	0.0245	8	2.66
PMMA	2	1	660	0.023	3940	23	0.0245	8	1.88
PMMA	4	2	660	0.023	3940	23	0.0245	8	1.20
FRP4	1	0	338	0.15	2130	10	0.106	10	5.18
FRP4	2	0	338	0.15	2130	10	0.106	10	3.07
FRP4	4	0	338	0.15	2130	10	0.106	10	1.61
FRP4	1	0.5	338	0.15	2130	10	0.106	10	5.72
FRP4	2	1	338	0.15	2130	10	0.106	10	3.45
FRP4	4	2	338	0.15	2130	10	0.106	10	1.86

It can be seen from Table 2 that safe separation distance for the PMMA varies from somewhat greater than the plan width for W_f = 4-m up to about 2.5 times the plan width for W_f = 1-m. There appears to be only a small effect of having a triangular cross section, instead of the rooflight being flush with the roof, as long as the height of the peak is less than half of the plan width. For the FRP rooflight material, safe separation distances are significantly greater than for the PMMA (ranging from 1.6 to 5.7 times the plan width), mainly because of the greater length of burning rooflight caused by the much higher flame spread rate for the FRP. In all cases, the calculated safe separation distance is greater than one-half the total flame height, so wind effects should be negligible.

Conclusions

1. A dimensionless flame acceleration factor (FAF) based on an existing simplified burnout model has been successfully used to evaluate the flame spread behavior of several thin polymeric rooflight materials.
2. A method for defining time to ignition, total burning duration and average chemical (actual) heat release rate per unit surface area has been developed to obtain the FAF from bench-scale tests.
3. Consistent with model predictions, flame spread remains within the exposure source in large-scale ASTM E 108 tests when the FAF values are less than zero or negative. When the FAF values are greater than zero, there is distinct flame spread beyond the exposure source, except for the case of some melting thermoplastic materials. This model does not take the melting phenomenon into account.
4. A method to calculate safe separation distance between rooflights has been developed for the case when flame spread over the rooflight from an exposure source is expected, based on FAF, but it is desired to prevent radiant ignition of adjacent, uninvolved rooflights.

Acknowledgments

The authors are grateful to the following persons who contributed to this project: Steve Ogden and Steve D'Aniello for conducting experimental work in Fire Propagation Apparatus (ASTM E 2058); Jeff Chaffee for conducting large-scale tests in ASTM E 108 Test Apparatus; Dr. Robert G. Bill, Jr, for valuable suggestions and support.

References

[1] Quintiere, J. G., Harkleroad, M., and Hasemi, Y., "Wall Flames and Implications for Upward Flame Spread," *Combustion Science and Technology*, Vol. 48, 1986, pp. 191-222.

[2] Mowrer, F. W. and Williamson, R. B., "Flame Spread Evaluation for Thin Interior Finish Materials," *Fire Safety Science – Proceedings of the Third International Symposium*, International Association of Fire Safety Science, Cox and Langford, Editors, 1991, pp. 689-698.

[3] Williams, F. W., Beyler, C. L., Hunt, S. P., and Iqbal, N., "Upward Flame Spread on Vertical Surfaces," Series Report, NRL/MR/6180-97-7908, Naval Research Lab., Washington, D.C., January 13, 1997, U.S. Department of Commerce, National Technical Information Service.

[4] Khan, M. M., and Alpert, R. L., "Evaluation of Flame Spread on Thin Polymeric Rooflight Materials," accepted for publication in the *Journal of Fire Sciences*.

[5] Cleary, T. G. and Quintiere, J. G., "A Framework for Utilizing Fire Property Tests," *Fire Safety Science – Proceedings of the Third International Symposium*, International Association of Fire Safety Science, Cox and Langford, Editors, 1991, pp. 647-656.

[6] NFPA 287, *Standard Test Methods for Measurement of Flammability of Materials in Cleanrooms Using a Fire Propagation Apparatus (FPA)*, NFPA International, P.O. Box 9101, Quincy, MA 02269-9101, USA, 2001.

[7] de Ris, J. L. and Orloff, L., "The Role of Buoyancy Direction and Radiation in Turbulent Diffusion Flames on Surface," *15th Symposium (International) on Combustion*, 1975, pp. 175, The Combustion Institute, Pittsburgh, PA.

[8] Dillon, S. E., Quintiere, J. G. and Kim, W. H., "Discussion of a Model and Correlation for the ISO 9705 Room-Corner Test," *Fire Safety Science – Proceedings of the Sixth International Symposium,* pp. 1015-1026.

[9] Markstein, G. H., "Scanning Radiometer Measurements of the Radiance Distribution in PMMA Pool Fires," *18th Symposium (International) on Combustion*, The Combustion Institute, Pittsburgh, PA, 1981, pp. 537-547.

[10] Markstein, G. H., "Radiative Properties of Plastics Fires," *17th Symposium (International) on Combustion*, The Combustion Institute, Pittsburgh, PA, 1978, pp. 1053-1062.

[11] Heskestad, G., "Fire Plumes, Flame Height and Air Entrainment," *The SFPE Handbook of Fire Protection Engineering*, Third Edition, P.J. DiNenno, editor, National Fire Protection Association, Quincy, MA, 2002, pp. 2-1 to 2-17.

Session II: Energy Conservation

Morton Sherman,[1] William Mullally, (Retired),[2] Jess Garcia,[2] and Diana Dobrez-Flores[2]

A Realistic Approach for Testing the Thermal Insulation Value of Closed-Cell Spray Polyurethane Foam Roof Insulation

---

Reference: Sherman, M., Mullally, W., Garcia, J., and Dobrez-Flores, D., "**A Realistic Approach for Testing the Thermal Insulation Value of Closed-Cell Spray Polyurethane Foam Roof Insulation,**" Roofing Research and Standards Development: 5th Volume, ASTM STP 1451, W. J. Rossiter and T. J. Wallace, Eds., ASTM International, West Conshohocken, PA, 2003.

Abstract: The thermal resistance characteristics of thicker closed-cell spray polyurethane foam insulation products appear to be understated when tested at one inch thickness as required in ASTM C1029-96 [1]. This results in excessive and costlier foam thicknesses being prescribed to meet a specifier's R-value requirements than if full product thickness conditioning before full product thermal resistance testing had been specified.

Confirmation of this concern was studied by the Building Envelope Committee of the Spray Polyurethane Foam Alliance, measuring full product thickness thermal resistances after the 180 days conditioning time. Three commercially marketed proprietary closed-cell spray polyurethane foam formulations containing HCFC 141b blowing agent were tested.

All three formulations affirmed the concept of measuring full product thickness thermal resistance after time conditioning. Increasingly thicker spray polyurethane foam specimens retained more of their initial thermal resistance values after the standard 180 days of time conditioning before testing, than if thermal resistance values measured on one inch thick spray foam specimens multiplied by the full product thickness, were used for extrapolating these thicker product thermal resistances, as done with homogeneous thermal insulations.

Keywords: apparent thermal conductivity, apparent thermal resistance, full product thickness, HCFC 141b blowing agent, normalized R-value, roof insulation, spray polyurethane foam, time conditioning

Introduction

Undocumented verbal reports from spray polyurethane foam users in walls, roofs and storage tank installations continue to suggest greater energy efficiency and fuel savings experiences than anticipated, particularly when compared with other thermal insulation materials used in similar applications.

In an attempt to explain or better understand these observations, an exploratory study of three thermal resistance sampling, conditioning, and testing modes, as detailed below, was initiated independently by consultMORTinc at a NVLAP (National Voluntary Laboratory Accreditation Program)-certified thermal testing laboratory on July I, 1996. Spray-applied closed-cell polyurethane foam specimens were produced on-site by a professional applicating contractor using full-size production equipment and a commercial polyurethane foam formulation.

---

[1] consultMORTinc., 6723 14th Avenue North, St. Petersburg, FL 33710-5405.

[2] Dow Chemical Co., 2050 North Broadway, Joliet, IL 60435-3187.

Thermal resistance tests were conducted at sequential time periods on foam specimens shelf-conditioned in a controlled/standard laboratory air environment for each of the three modes studied. Summarizing the study results presented orally at the spray polyurethane foam industry's "Spray Foam '98" conference held in Sparks, Nevada in February, 1998:

(A)Time conditioning 3.5 in. thick spray foam specimens, followed by the removal of their top and bottom surface layers, permitted thermal resistance testing as one inch thick core foam specimens. These specimens exhibited essentially NO thermal resistance decrease over twelve months conditioning time. An ASTM C 518 [1] heat flow meter apparatus was used.

(B)Time conditioning one-inch thick spray polyurethane foam specimens and sequentially testing for thermal resistance against time DID exhibit logarithmic reductions in R-value per inch thickness as expected over twelve months. An ASTM C 518 heat flow meter apparatus was used.

(C)Time conditioning a 3.5 in. thick spray polyurethane foam specimen and its sequential testing against time, exhibited very slowly declining thermal resistance values over twelve months conditioning time, much like a closed-cell foam material protected from air gas permeation. An ASTM C 236 guarded hot box apparatus [1] was used.

It was concluded that the one-inch thick foam specimen preparation and dimensions specified in ASTM C 1 029-90 are in need of re-examination and discussion by those having jurisdiction over this document The thermal resistance characteristics of thicker closed-cell spray polyurethane foam insulation products appear to be understated when tested as required in ASTM C 1029-96. This results in excessive and costlier foam thicknesses to meet a specifier's R-value requirements than if full product thickness conditioning before full product thickness testing had been specified.

Confirmation of this concept was initiated in March, 2000, with a study by the Building Envelope Committee of the Spray Polyurethane Foam Alliance (SPFA).[3] Objective: to measure the full-product-thickness thermal resistance of three commercially marketed proprietary closed-cell polyurethane foam formulations after following the industry's standard 180 day laboratory conditioning convention for time conditioning before thermal testing.

It is well-known that the thermal resistance characteristics of homogeneous thermal insulations in this 2–3 lbs/cf product density range have a direct and linear relationship with thickness. However, when the outer surface areas of a closed-cell foam plastic thermal insulation receive air (gas) permeation, which dilutes and reduces the insulation characteristics of these outer strata, the resultant composite foam plastic material (air-permeated outer strata and non-permeated inner core foam strata) can be considered to be nonhomogeneous. With core cellular strata having an undiluted cell gas composition and thermal resistance, these core foam strata remain as yet unaffected by R-value decay.

Spray polyurethane foam is manufactured on-site. Successive traverse spray applications or "lifts" are utilized to achieve the required applied material thickness. Each traverse results in a significant top layer of denser closed-cell "skin" between spray foam

[3] 1300 Wilson Boulevard, Arlington, VA 22209.

applications as the final applied thickness of foam insulation is achieved. This reality also prevents spray polyurethane foam from being termed "homogeneous." These traverse foam thicknesses and skin layers are neither of fixed dimensions nor frequency within a particular spray foam application. The presence of "skins" might also act to retard air permeation. The thermal testing of end-use product thicknesses would appear to be a most appropriate and understandable testing mode under these circumstances.

Experimental Details

Three suppliers of proprietary spray polyurethane foam systems offered to provide spray foam materials for this SPFA study. All three chemical formulations contained HCFC 141b blowing agent. Other formulation details remained unknown for the purposes of this study, since their compositions remain proprietary with each producer.

Four foot by four foot panel specimens ranging from one inch to four inches in spray foam thickness, were produced as described previously. Panels were shipped without delay to the Dow Chemical Company's NVLAP-accredited testing laboratory in Joliet, Illinois.

Conditioning was conducted against calendar time in a controlled laboratory environment for 180 days. Thermal resistance testing of three common lot spray foam specimens was then conducted on each available lot thickness in accordance with ASTM C 518-98.

Data measurements are tabulated and reported in Tables 1–3 for materials provided by Suppliers A, B, and C.

Specimen thicknesses and measured R-values were normalized so that R-values of similar thickness products would be directly comparable and reported in Tables 4–6.

The following relationships were derived from

$$k = t / R^{*} \quad (1)$$

Where k = apparent thermal conductivity
t = spray foam thickness
R = apparent thermal resistance

Assuming, k[measured] = k[normalized] for these small thickness adjustments, then (2)

$$k[\text{measured}] = t[\text{measured}] / R[\text{measured}] \quad (3)$$

$$k[\text{normalized}] = t[\text{normalized}] / R[\text{normalized}] \quad (4)$$

$$R[\text{normalized}] = R[\text{measured}] \times \frac{t[\text{normalized}]}{t[\text{measured}]} \quad (5)$$

* R is expressed herein in inch-pound units ($h*ft^2*F/Btu$).

TABLE 1—*Laboratory data from Supplier A spray polyurethane foam material.*

Thickness Inches	Triplicate Specimen	Apparent k	Apparent R	Averaged R	Averaged Normalized R	Density pcf
Initial 1.000"						
1.030	1	0.128	8.047			1.98
1.033	2	0.128	8.070			1.95
1.029	3	0.127	8.102	8.070	7.8	1.96
Conditioned 1.000"						
1.030	1	0.165	6.242			1.94
1.031	2	0.166	6.211			1.92
1.027	3	0.166	6.187	6.213	6.0	1.97
Initial 2.000"						
1.931	1	0.134	14.410			1.96
1.929	2	0.129	14.953			2.01
1.926	3	0.127	15.165	14.843	15.4	1.99
Conditioned 2.000"						
1.951	1	0.162	12.043			1.98
1.948	2	0.159	12.252			2.02
1.949	3	0.154	12.656	12.317	12.6	1.98
Initial 3.000"						
2.895	1	0.133	21.767			1.97
2.881	2	0.133	21.662			1.96
2.895	3	0.132	21.932	21.787	22.6	2.01
Conditioned 3.000"						
2.914	1	0.156	18.679			2.03
2.905	2	0.155	18.742			2.01
2.921	3	0.152	19.217	18.879	19.4	2.03
Initial 4.000"						
3.957	1	0.135	29.311			2.01
3.991	2	0.134	29.784			2.02
3.986	3	0.133	29.970	29.688	29.9	2.03
Conditioned 4.000"						
4.004	1	0.152	26.342			2.06
4.022	2	0.152	26.461			2.11
4.037	3	0.149	27.094	26.632	26.5	2.07

NOTE: Laboratory test data by Dow Chemical Co./Joliet, Illinois Laboratory. Normalization calculations by consultMORTinc, St. Petersburg, Florida.

TABLE 2—*Laboratory data from Supplier B spray polyurethane foam material.*

Thickness Inches	Triplicate Specimen	Apparent k	Apparent R	Averaged R	Averaged Normalized R	Density pcf
Initial 1.000"						
1.142	1	0.146	7.822			2.73
1.028	2	0.144	7.139			2.42
1.152	3	0.146	7.890	7.617	6.9	2.99
Conditioned 1.000"						
1.144	1	0.183	6.251			2.74
1.034	2	0.186	5.559			2.40
1.141	3	0.174	6.557	6.122	5.5	3.03
Initial 2.000"						
2.152	1	0.148	14.541			3.40
2.024	2	0.146	13.863			2.78
2.051	3	0.145	14.145	14.183	13.7	2.66
Conditioned 2.000"						
2.121	1	0.160	13.256			3.51
2.054	2	0.172	11.942			2.83
2.070	3	0.167	12.395	12.531	12.0	2.69
Initial 3.000"						
3.038	1	0.139	21.856			2.83
3.054	2	0.140	21.814			3.09
3.003	3	0.142	21.148	21.606	21.4	3.24
Conditioned 3.000"						
3.048	1	0.151	20.815			2.83
3.047	2	0.152	20.046			3.11
2.986	3	0.156	19.141	19.791	19.6	3.29
Initial 4.000"						
4.048	1	0.147	27.537			3.15
4.025	2	0.138	29.167			3.05
3.996	3	0.138	28.957	28.554	28.4	2.77
Conditioned 4.000"						
4.056	1	0.149	27.221			3.15
4.011	2	0.148	27.101			3.06
4.013	3	0.142	28.261	27.528	27.4	2.78

NOTE: Laboratory test data by Dow Chemical Co./Joliet, Illinois Laboratory. Normalization calculations by consultMORTinc, St. Petersburg, Florida.

TABLE 3—*Laboratory data from Supplier C spray polyurethane foam material.*

Thickness Inches	Triplicate Specimen	Apparent k	Apparent R	Averaged R	Averaged Normalized R	Density pcf
Initial 1.000"						
1.081	1	0.134	8.067			2.60
1.038	2	0.134	7.746			2.67
1.020	3	0.135	7.556	7.790	7.4	2.82
Conditioned 1.000"						
1.083	1	0.162	6.685			2.59
1.041	2	0.164	6.348			2.66
1.023	3	0.162	6.315	6.449	6.2	2.80
Initial 2.000"						
2.043	1	0.137	14.912			2.50
2.095	2	0.138	15.181			2.69
2.059	3	0.136	15.140	15.078	14.6	2.50
Conditioned 2.000"						
2.052	1	0.154	13.325			2.49
2.100	2	0.151	13.907			2.69
2.066	3	0.153	13.503	13.578	13.1	2.50
Initial 3.000"						
2.968	1	0.138	21.507			2.54
2.948	2	0.138	21.362			2.50
3.055	3	0.140	21.843	21.571	21.6	2.42
Conditioned 3.000"						
2.980	1	0.147	20.272			2.53
2.962	2	0.147	20.150			2.49
3.071	3	0.148	20.750	20.391	20.4	2.43
Initial 4.000"[a]						
N/A						
N/A						
N/A						
Conditioned 4.000"[a]						
N/A						
N/A						
N/A						

[a] Four inch thickness specimens not provided by Supplier C.
NOTE: Laboratory test data by Dow Chemical Co./Joliet, Illinois Laboratory. Normalization calculations by consultMORTinc, St. Petersburg, Florida.

TABLE 4—*Spray foam from Supplier A.*

Normalized Thickness	Average Initial R-value	Average Conditioned R-value
1.000 in.	7.8 R	6.0 R
2.000 in.	15.4 R	12.6 R
3.000 in.	22.6 R	19.4 R
4.000 in.	29.9 R	26.5 R

TABLE 5—*Spray foam from Supplier B.*

Normalized Thickness	Average Initial R-value	Average Conditioned R-value
1.000 in.	6.9 R	5.5 R
2.000 in.	13.7 R	12.0 R
3.000 in.	21.4 R	19.6 R
4.000 in.	28.4 R	27.4 R

TABLE 6—*Spray foam from Supplier C.*

Normalized Thickness	Average Initial R-value	Average Conditioned R-value
1.000 in.	7.4 R	6.2 R
2.000 in.	14.6 R	13.1R
3.000 in.	21.6 R	20.4R
4.000 in.	N/A	N/A

Conclusions

Table 7 compares the R-values extrapolated for thicker spray foam products using the product thickness multiplier convention for homogeneous materials as mentioned above, versus full product thickness R-values as measured in this study. Now that affordable and accurate thermal resistance testing equipment is widely available for readily testing full thickness thermal insulation materials to as much as six inches or more, a return to full thickness testing after time conditioning deserves serious consideration by both producers and users in updating Material Specification C 1029 without undue delay.

TABLE 7—*Product R-values at four product thicknesses; extrapolated R-values (homogeneous convention) versus measured full thickness R-values.*

Normalized Foam thickness , in.	Extrapolated R-value, $h*ft^2*F/Btu$	Full Thickness Measured R-value, $h*ft^2*F/Btu$	Percent Difference
1	5.9 R	5.9 R	0
2	11.8 R	12.6 R	7
3	17.7 R	19.8 R	12
4	23.6 R	27.0 R	14

As shown in Table 8, all three spray foam polyurethane thermal insulation formulations confirmed the validity of full product thickness sampling and conditioning before full product thickness testing for thermal resistance. All three demonstrated that increasingly thicker specimens lost less of their initial thermal resistance values after time conditioning (see Tables 7 and 8).

TABLE 8—*Average R-values. Three lots of spray polyurethane foam at normalized thicknesses.*

Thickness	Initial R-value	Average	180 Day Conditioned R-value	Average	% of Initial R-value Retained After Aging
1.000 in.	7.8		6.0		
	6.9		5.5		
	7.4		6.2		
		7.4 R		5.9 R	80
2.000 in.	15.4		12.6		
	13.7		12.0		
	14.6		13.1		
		14.6 R		12.6 R	86
3.000 in.	22.6		19.4		
	21.4		19.6		
	21.6		20.4		
		21.9 R		19.8 R	90
4.000 in.	29.9		26.5		
	28.4		27.4		
	N/A		N/A		
		29.2 R		27.0 R	93

It becomes increasingly apparent that full product thickness conditioning followed by full product thickness testing by ASTM C 518 or ASTM C 236 test methods is a more realistic characterization mode for spray polyurethane foam insulation materials, and perhaps some other closed-cell foam plastic insulation materials as well.

Table 9 is a tabulation of interpolated time conditioned R-values for thickness fractions between averaged thickness data produced in this study of three proprietary spray foam formulations. This demonstrates how Task Group C 1029 of Subcommittee C16.22 might develop such a table for inclusion in a future revision of ASTM C 1029.

TABLE 9—*Interpolation of fractional thickness R-values from conditioned full-thickness averages developed in this three formulations study (see Table 8).*

Normalized Thickness, in.	Average R-values	Interpolated R-values for Fractional Thicknesses
1	5.9 R	
1.25		7.6 R
1.5		9.3 R
1.75		10.9 R
2	12.6 R	
2.25		14.4 R
2.5		16.2 R
2.75		18.0 R
3	19.8 R	
3.25		21.6 R
3.5		23.4 R
3.75		25.2 R
4	27.0 R	

Acknowledgment

The authors and the spray polyurethane industry are very much indebted to the Dow Chemical Company for providing the resources of their laboratory, equipment access, and personnel time for the conduct of this investigation and the presentation of this paper, without hesitation or financial compensation. We believe that this is a prime example of responsible product stewardship by a leading chemicals supplier corporation to the industry it services. In addition, consultMORTinc expresses appreciation for the cooperation and technical professionalism of the employees of the Dow Chemical Company/Joliet.

References

[1] *Annual Book of ASTM Standards*, ASTM International, West Conshohocken, PA.

Joseph W. Mellott II[1] *and Donald C. Portfolio*[2]

The Effect of Reflective Roof Coatings on the Durability of Roofing Systems

REFERENCE: Mellott, J. W. and Portfolio, D. C., "**The Effect of Reflective Roof Coatings on Durability of Roofing Systems,**" *Roofing Research and Standards Development: Fifth Symposium, ASTM STP 1451,* W. J. Rossiter and T. J. Wallace, Eds., ASTM International, West Conshohocken, PA, 2003.

ABSTRACT: Based on the knowledge obtained from the Roof Coatings Manufacturers Association (RCMA) sponsored Oak Ridge National Laboratory studies completed in 2000, the RCMA is launching a new long-term durability study. In the original study, preliminary data indicated that reflective coatings retarded the degradation of built-up roofing systems. This part of the study was minimal in scope and incidental to the original test program. As a result this new, more inclusive study is being undertaken.

In the new study the effect of seven reflective coatings and a control will be determined over four different roofing membranes. The purpose of the study is to determine the long-term effect of the coatings on performance properties of the roofing membrane systems by the applied reflective coatings. The coatings used in the study will be selected from commercially available products. The roofing systems will be exposed in three different geographic regions: the midwest, the southeast and the southwest. The membranes will include a granule surfaced SBS membrane system, a smooth APP membrane system, a standard built-up roof system, and a black, non-reinforced EPDM membrane.

The reflective roof coatings selected include two types of fibered aluminum coating, one white styrene acrylic elastomeric coating, one white acrylic elastomeric coating, one white solvent based elastomeric coating, one water-based aluminum coating, and a standard fibered black asphalt emulsion coating. The systems will be periodically tested throughout the aging process for a group of physical properties related to performance.

Concurrent with the outdoor exposure, accelerated aging will be conducted using two conventional weathering techniques. The objective in this part of the study is to determine if there is a correlation between accelerated aging and in-field exposure. The subsequent effect of accelerated weathering on changes in reflectance and emissivity will be determined. The accelerated data will be compared to existing 3-year in-field data to determine if there is a correlation between the accelerated weathering and real time weathering.

KEYWORDS: reflective, roof, coatings, durability, aluminum, acrylic, asphalt

Introduction

In 1997 the Roof Coating Manufacturers Association (RCMA) embarked on a research program with Oak Ridge National Laboratory (ORNL) to determine the effect of reflective coatings on heat transfer through a small scale roofing system over time using both aluminum pigmented asphalt based coatings and white elastomeric roof reflective coatings. The solar reflectance and infrared emittance of these coatings would also be

[1] Vice President, Momentum Technologies, Inc., Uniontown, OH 44685.

[2] Vice President, PRI Asphalt Technologies, Inc., Tampa, FL 33610.

monitored along with resulting roofing membrane temperatures. These measured properties impact the energy consumption used for heating and cooling depending on geographic location.

RCMA realized that the relative real-time weathered performance of these types of coatings manufactured by the members of the organization should be quantitatively determined. Oak Ridge was interested in the program as they wanted to determine the impact of reflective coatings on the heat and cooling loads in commercial buildings.

The Oak Ridge National Laboratory Work

The RCMA Technical Committee was charged by the organization with determining which coatings would be used in the Oak Ridge study. The organization decided to supply ORNL with the following types of coatings:

1. Aluminum Roof Coating (Non-Fibrated)
2. Aluminum Roof Coating (Fibrated – Good Grade)
3. Aluminum Roof Coating (Fibrated – Medium Grade)
4. Aluminum Roof Coating (Fibrated – Economy Grade)
5. White Elastomeric Coating (Acrylic)
6. White Elastomeric Coating (Styrene-Acrylic)

RCMA Coating Formulations

The four asphalt cutback based aluminum roof coatings were generic formulations developed by the RCMA Technical Committee and the two white elastomeric formulations were standard starting formulations developed by the suppliers of the latices used in the manufacture of the coatings. It was realized at the onset of the program that generic formulations for the Aluminum products would not represent the best products that were, at the time, commercially available, but would give a general and relative indication of the three year performance of aluminum roof coating products. The same was recognized for the white coatings which were "starting" formulations developed by the latex suppliers. The coatings were all prepared in accordance with the formulations mentioned and sent to ORNL for application.

Roofing Membrane

The roofing membrane chosen by ORNL and RCMA was a 4 ply built-up roof assembly. The assembly consisted of a nailed Standard Specification for Asphalt-Coated Glass Fiber Base Sheet Used in Roofing (ASTM D 4601), Type II base sheet and three plies of Standard Specification for Asphalt Glass Felt Used in Roofing and Waterproofing (ASTM D 2178), Type IV asphalt glass felt adhered with Standard Specification for Asphalt Used in Roofing (ASTM D 312), Type IV asphalt. The assemblies were applied to the test deck as in an actual roofing situation. The base sheet was nailed to the plywood and then three plies of glass felt were applied in shingle fashion to the base sheet.

Coating Procedure

The membrane was coated with each of the six materials in accordance with good roofing application practice. The aluminum coatings were applied in two coats. The fibrated materials were applied at approximately 1.5 – 2 gallons per square and the non-fibrated applied at 1 gallon per square. The white coatings were also applied in two coats at a rate of 1.5 gallons per square. One of the panels was left without coating as a basis for comparison or the control. These coverage rates were considered average for the specific type of coating.

Test Protocol

As mentioned above, the primary purpose of this program for the RCMA was to determine the maintenance of relative reflectance performance of these generic coatings in real time. As an after thought, since the opportunity presented itself, it was also decided to do some simple testing on the roof cores before and after real time weathering to determine if there was an effect on the performance of primarily the asphalt component in the roof membrane by the various coatings. It has been postulated that the aging rate of asphalt in a roofing membrane is related to the temperature of the membrane. This project offered the opportunity to test this hypothesis in real time since it was known that the temperature of the membrane would be related to the reflectance of the surface of the membrane, all other things being equal, which they were in this program.

ORNL measured the solar reflectance of these coatings at intervals throughout a three year period along with other heat transfer related properties of the system. These results were reported in other publications [1]. Of interest in the smaller informal part of the project was the initial and final reflectance of the coatings used in the study. The solar reflectance was measured using a Devices and Services portable solar spectrum reflectometer. Results determined with this reflectometer agree well with reflectance determined using ASTM Standard Test Method for Solar Absorptance, Reflectance, and Transmittance of Materials Using Integrating Spheres (E 903) [2].

Along with the properties determined by ORNL, the RCMA decided to determine the following properties on the built-up membrane systems with various coatings at the beginning and end of the study:

Interply asphalt analysis

Softening Point:	ASTM Standard Test Method for Softening Point of Bitumen (Ring-and-Ball Apparatus) (D 36)
Penetration at 25°C:	ASTM Standard Test Method for Penetration of Bituminous Materials (D 5)

Roofing Membrane

Load at -17.8°C at Failure:	ASTM Standard Practice for Testing Load-Strain Properties of Roofing Membranes (D 2523)
Strain at -17.8°C at Failure:	(ASTM D 2523)

The interply asphalt was removed from the cores by freezing and then splitting the cores and then carefully scraping the interply asphalt from the core. Sufficient asphalt was collected to fill three softening point rings. Two of the rings were used to determine the softening point and the third ring was used for penetration testing. While the penetration testing was not strictly in accordance with ASTM D 5, due to the quantity of asphalt used and specimen container, for low penetration materials, this deviation has little effect on the result and in any event has no effect on the relative differences noted in the penetration results.

The load strain properties were determined in accordance with ASTM D 2523 at -17.8°C. Three specimens in each membrane direction comprised the sample set. Since the membranes were applied shingle fashion as they would be in the field, the load strain properties are confounded by the number of plies and/or laps present in the area of the specimen cut from the core. It was not until after the membranes were constructed that this program was developed.

Results and Discussion

Compiled below are the results of the testing of these membrane systems including the ORNL reflectance measurements and the physical properties of the asphalt and the stress-strain analysis of the BUR cores completed by RCMA. The properties were determined initially and after three years of exposure at ORNL in Tennessee. The initial values obtained for the un-coated control were used as the initial properties for all of the specimens.

Asphalt Analysis

TABLE 1 - *Control specimen (Uncoated BUR).*

Coating Type	$Reflectance_i$	$Reflectance_3$	SP_i, °C	SP_3, °C	Pen_i, dmm	Pen_3, dmm
No Coating (Control)	0.06	0.06	104	131	12	4

Note: i = Initial, 3 = three year aged result

TABLE 2 – *Coated BUR specimens.*

Coating Type	$Reflectance_i$	$Reflectance_3$ Pre-wash	$Reflectance_3$ Washed	SP_3, °C	Pen_3@25°C, dmm
Acrylic White	0.83	0.52	0.61	114	7
Styrene Acrylic White	0.86	0.52	0.70	109	8
Aluminum Fib. – Good	0.69	0.41	0.55	122	5
Aluminum Fib.– Medium	0.64	0.34	0.44	118	6
Aluminum Fib. - Economy	0.54	0.27	0.28	121	5
Aluminum N. F.	0.71	0.39	0.49	124	4

Note: Fib. = Fibrated, NF = Non-fibrated

It is evident from these results, that the use of reflective coatings has a measurable effect on the change in softening point and penetration of interply asphalt over the three year period of natural weathering. Generally, the higher the initial and three year reflectance value, the smaller the change in the properties of the interply asphalt used to construct the membranes. The differences in asphalt properties between the aluminum coated BUR samples are relatively small and clustered. There appears to be a measurable difference in the changes in asphalt properties between the aluminum and white coated BUR samples and the un-coated control.

The increase in softening point and decrease in penetration is a result of the aging of the asphalt. In this case, the properties of the interply asphalt, which was not exposed to the sun directly, is primarily a result of heat. These changes are expected. While the reflective coatings certainly slow the pace of aging, properly designed asphalt roofing systems have a long history of excellent performance without these coatings. It is expected that this type of roofing system coupled with a reflective coating which lowers the average membrane temperature [*3*] has the potential to increase the life of this type of roofing system.

It should be noted that the aged reflectance values were obtained for the specimens for two different conditions: pre-washed and washed. The pre-washed condition indicates the condition of the condition of the coating in terms of reflectance at the end of the three year exposure period. The washed condition indicates the potential reflectance of the coating if the membrane were washed using mild conditions.

Stress Strain Analysis of the BUR System

TABLE 3 - *Uncoated BUR control.*

Property	Control, Initial	Control, 3 Year
Load at failure @ -17.8°C, kN/m		
MD	70.3	86.3
CMD	58.1	68.3
Strain at failure @ -17.8°C, %		
MD	3.3	2.9
CMD	3.0	3.2
Number of Plies	5.6	4.5

TABLE 4 - White coated BUR.

Property	Acrylic White	Styrene Acrylic
Load at failure @ -17.8°C, kN/m		
MD	69.5	52.0
CMD	60.0	50.8
Strain at failure @ -17.8°C, %		
MD	2.8	2.5
CMD	2.7	2.4
Number of Plies	5.0	4.4

TABLE 5 - *Aluminum coated BUR.*

Property	Aluminum Fib Good	Aluminum Fib Medium	Aluminum Fib Economy	Aluminum NF
Load at failure @ -17.8°C, kN/m				
MD	69.5	75.8	70.5	66.5
CMD	48.5	49.5	61.2	59.5
Strain at failure @ -17.8°C, %				
MD	3.0	3.4	3.4	3.5
CMD	2.9	3.2	3.1	3.0
Number of Plies	4.3	4.2	5.0	5.0

As mentioned earlier, the results of the stress-strain analysis at -17.8°C are somewhat confounded by the method used to construct the membranes which resulted in varying number of plies, depending on specimen location within the sample. Bearing this in mind, along with the fact that the tensile of a membrane is dominated by the reinforcement, there is some evidence that the reflective roof coating resulted, overall, in lower three year aged tensile values when compared to the control taking into account the number of plies in the specific specimens tested. The increase in tensile is related to the degree of hardening of the sample age.

New RCMA Program

Building on the data established in the preliminary Oak Ridge study, the Roof Coating Manufacturers Association developed a protocol for a more extensive examination of the effect of roof coatings on the long term durability of roof systems.

Coating Selection

Cross sections of commercially available roof coatings were selected for the new study. The coatings selected were commercially available products considered to represent the middle tier of the family of products they were chosen to represent. The product families selected were as follows:

High Softening Point Asphalt Cutback Based Aluminum Roof Coating (Fibered)
Low Softening Point Asphalt Cutback Based Aluminum Roof Coating (Fibered)
Water Based Aluminum Roof Coating
Solvent Based White Elastomeric Roof Coating
Styrene Acrylic Based White Roof Coating
Acrylic Based White Roof Coating
Asphalt Emulsion

The specific roof coatings were chosen by committee and samples were requested, at minimum, from two commercially available sources. The double blind system was developed to protect the anonymity of the material sources.

RCMA concluded that in the earlier study, reluctance to provide actual product formulation resulted in possible sub standard product formulations. By selecting commercially available products, better performance is expected so that a clearer demonstration of the effect of the coating on the life of the membrane can be demonstrated. There is also the long term possibility to create a reference to the known field performance of the products.

In all cases the coatings will be applied in compliance with the manufacturers' recommendations for each of the specific substrates. All appropriate primers and pre-application surface preparations will be used. For certain substrates in the study, recommendations are not available. In these cases the coatings will be applied without primer to the specific substrate.

Roofing Systems

To develop data for a large selection of membrane systems, four systems were chosen to establish the widest market coverage. The systems selected were an asphalt built-up roof (BUR), two polymer modified bituminous membrane systems including styrene-butadiene-styrene modified (SBS), and atactic polypropylene modified (APP), and black EPDM single ply membrane. As in the previous study sample decks were prepared for subsequent coating. The systems constructed are as follows:

Membrane Type: BUR
Deck: Plywood
Insulation: 2 in. Isocyanurate / ½ in. Fiberboard (Mechanically Attached)
Ply: G2 Base (Hot Mopped in Type IV Asphalt, 25 lb./100 ft^2)
2 – G1 Felts (Hot Mopped in Type IV Asphalt, 25lb./100 ft^2)
Surface Membrane: Type IV Asphalt Glaze Coat (50 lb./100 ft^2)

Membrane Type: SBS
Deck: Plywood

Insulation: 2 in. Isocyanurate / ½ in. Fiberboard (Mechanically Attached)
Ply: G2 Base (Hot Mopped in Type IV Asphalt, 25lbs.)
Surface Membrane: Granule SBS Membrane (Hot Mopped in Type IV Asphalt)

Membrane Type: APP
Deck: Plywood
Insulation: 2 in. Isocyanurate / ½ in. Fiberboard (Mechanically Attached)
Ply: G2 Base (Hot Mopped in Type IV Asphalt, 25lbs.)
Surface Membrane: Smooth APP Membrane (Torched)

Membrane Type: EPDM
Deck: Plywood
Insulation: 2 in. Isocyanurate / ½ in. Fiberboard (Mechanically Attached)
Surface Membrane: Black Unreinforced EPDM (Fully Adhered)

The systems were constructed for subsequent testing. Ply sheets were applied in one-over-one fashion so that the core extractions made throughout the aging cycle will result in an equivalent ply count at each sampling. Each of the deck systems will be aged for a period of six months prior to coating application.

Aging Locations

An improvement over the last study is using multiple aging sites for the roofing systems. In the previous, study the deck systems were aged in one region, considered to be a moderate climate. In the new study three sites were selected. The sites were selected to provide a cross section of North American climatic conditions. Each of the previously described decks was constructed in triplicate in North East Ohio. One set of the deck systems will be aged in the North East Ohio Region (Industrial Midwest Site) another in the South East (Tampa, Florida - Subtropical) and the third in the South West (Phoenix, Arizona - Desert). Periodically, one foot square test samples will be removed from the decks. Each of the test samples will be analyzed for a battery of tests to establish the degree of aging of the substrate system.

To establish initial data small decks of each of the described types were produced along with the larger test decks. These decks once coated will be sacrificed to perform the battery of tests as described to develop day zero data.

Test Protocol

The new test protocol can be divided into two distinct areas, chemical and physical analysis. The chemical analysis of the asphalt based systems will identify specific chemical differences between the coated and un-coated systems upon aging. This test has been selected to yield information on basic composition changes in the asphalt as a result of aging.

The physical analysis group was established to provide a two-fold level of information. Data has been previously established concerning physical changes as a result of aging in roofing system. The physical tests will first attempt to establish these differences then develop correlation to durability. Further, the physical tests will also establish actual performance changes in the roofing systems. This study will attempt to show the effect on long term performance of the roof system of the use of roof coatings.

Asphalt Testing

The chemical test to be utilized is the Iatroscan. The Iatroscan separates asphalt into compositional components through the use of thin layer chromatography and solvent elution with flame ionization detection. The Iatroscan will be used to establish the base composition of the key components of asphalt. These key components change during the ageing process. Asphalt consists primarily of saturates, asphaltenes, resins, and aromatics. An Iatroscan identifies the amount of each of these components in asphalt. By comparing the original known composition of the Type IV asphalt, SBS compounds, and APP compounds subsequent component analyses can be conducted to determine the difference between the original values and the aged coated and non-coated component composition of the asphalt. This test is routinely used as a tool to identify changes in asphalt and polymer modified asphalt composition.

Dynamic shear rheometry (DSR) will be used to characterize complex viscosity properties of the asphalt. By conducting a variety of strain, frequency, and temperature sweeps, the temperature susceptibility of the asphalt along with other complex shear and temperature related complex rheological properties can be determined. By determining changes in these complex viscosity properties in the asphalt, some correlation to performance may be evident. DSR analysis is still an experimental method when used in the analysis of roofing asphalts, however is well defined and common in the analysis of paving asphalts. Specific protocol will be established throughout the process of analysis to yield the most reliable data. This effort in itself will prove to be extremely valuable in the area of rheological analysis of asphalt.

Other Testing

A small battery of physical tests will be used to establish change in performance and loss in function of the roof system. The physical tests for the asphalt based systems include: tensile/elongation, softening point, and penetration. The EPDM system will be analyzed for tensile/elongation, cold bend, and thickness throughout the ageing process. It is believed that this series of tests will best characterize changes in the roofing systems and attempt to establish differences between coated and un-coated parts of the roof system.

Solar Reflectance, Infrared Emittance, and Accelerated Aging

During the aging process of the decks, the coatings will be analyzed periodically for solar reflectance and infrared emittance. The intent is to examine the decks for at minimum 3 years. Data will be gathered through this aging cycle. Direct comparison can be established between regional aging sites. Comparisons between material performance can also be established.

Further and simultaneous to this effort, samples will be prepared on aluminum panels for accelerated aging. Accelerated aging equipment will include both Xenon Arc and QUV type equipment. The protocol utilized will be both ASTM and non-ASTM procedures. The effort in this portion of the study is to establish correlation between in-field and accelerated aging.

Conclusion

The ORNL project built the foundation for the forward RCMA study. The initial findings indicated that the use of reflective coatings appears to have a life cycle performance benefit for asphalt built-up roofing. By expanding the study to include more substrates, locations, coatings, and test protocols, a better estimation of this life-cycle benefit can be established. Further, it will be a valuable tool if a correlation between accelerated aging methods and in-field aging product performance can be established.

Acknowledgments

Thanks to Tremco, Inc., Cleveland, OH for conducting the analysis of the original control sample and white coating preparation, Karnak Corp., Clark, NJ for preparing the aluminum coatings, and PRI Asphalt Technologies, Inc., Tampa, FL for conducting the analyses of the aged samples.

Thanks to Momentum Technologies, Inc., Uniontown, OH and Simon Roofing, Youngstown, OH for construction of the decks for the new study.

Reference

[1] Wilkes, K. E., Petrie, T. W., Atchley, J. A. and Childes, P. W., "Roof Heating and Cooling Loads in Various Climates for the Range of Solar Reflectances and Infrared Emittances Observed for Weathered Coatings, "*Proceedings 2000 ACEEE Summer Study on Energy Efficiency in Buildings: Efficiency and Sustainability,* American Council for an Energy Efficient Economy, Washington D. C., 2000.

[2] Petrie, T. W., Dejarlais, A. O., Robertson, R. H. and Parker, D. S., "Comparison of Techniques for in-situ, Non-Damaging Measurements of Solar Reflectance of Low-Slope Roof Membranes," *International Journal of Thermophysics,* 2000.

[3] Opcit, Wilkes, et al

Session III: Durability

Carl C. De Leon,[1] *Patrick W. Diebel,*[1] *Ana H. Delgado,*[2] *and Ralph M. Paroli*[2]

A Comparison of the Glass Transition Temperatures of Artificially Aged and Field-Aged Roof Membranes Through the Use of Dynamic Mechanical Analysis

REFERENCE: De Leon, C. C., Diebel, P. W., Delgado, A. H., and Paroli, R. M., **"A Comparison of the Glass Transition Temperatures of Artificially Aged and Field-Aged Roof Membranes Through the Use of Dynamic Mechanical Analysis,"** *Roofing Research and Standards Development: Fifth Volume, ASTM STP 1451*, T. J. Wallace and W. J. Rossiter, Eds., ASTM International, West Conshohocken, PA, 2003.

ABSTRACT: The authors studied the effect of heat and ultraviolet (UV) exposure as well as natural weathering on the glass transition temperature of PVC roof membranes. The glass transition temperature of the unexposed specimens and those exposed to heat, artificial (UV) and natural weathering obtained from the E' and E" curves, as per ASTM Practice for Dynamic Mechanical Analysis and Thermogravimetry of Roofing and Waterproofing Membrane Material (D 6382) and ASTM Test Method for Plastics: Dynamic Mechanical Properties: In Flexure (Dual Cantilever Beam) (D 5418), range from –48 °C to –33 °C and from –39 °C to –28 °C, respectively. The ΔT_g values obtained from the E' curve varied between –6 °C and 4 °C and those from the E" curve varied between –4 °C and 9 °C. The results indicate that some naturally exposed membranes have similar ΔT_gs to some of the artificially aged samples.

KEYWORDS: dynamic mechanical analysis, glass transition temperature, roof membranes, poly(vinyl chloride).

Introduction

Much effort has been expended on the use of thermal analyses for use as predictive tools to assess the service life of roof membranes [*1-23*]. Thermoanalytical techniques can be used to monitor a wide array of material characteristics. Some of the applications include enthalpy, weight-loss, thermal stability, coefficient of thermal expansion (CTE) and the glass transition temperature (T_g). T_g is an important material characteristic since below T_g the material is rigid and hard while above T_g the material is flexible. Thus, T_g could be useful in evaluating the cold temperature performance of roof membranes [*14*].

Many physical properties can be studied by the various thermoanalytical techniques, including thermogravimetry (TG), differential thermal analysis (DTA), differential scanning calorimetry (DSC), thermomechanical analysis (TMA), and dynamic mechanical analysis (DMA). One property, which appears to show promise, is the glass transition temperature (T_g), and the related changes in T_g (ΔT_g) induced by laboratory and field-aging. This may be useful in assessing properties such as low temperature performance and service life. DMA is an excellent technique to monitor T_g.

Dynamic mechanical analysis provides information on mechanical properties. The

[1]Canadian General-Tower Limited, 52 Middleton St., P.O. Box 160, Cambridge, ON, Canada N1R 5T6.

[2]National Research Council of Canada, Institute for Research in Construction, Ottawa, ON, Canada K1A 0R6.

DMA technique measures a dynamic stress-strain relationship for a viscoelastic material.
Either a flexural or tensile load can be applied at predetermined frequency; modulus data are monitored.

In conducting a DMA experiment, two components of modulus can be obtained by resolving the stress-strain components:

$$\sigma = \varepsilon_o E'\sin(\omega t) + \varepsilon_o E''\cos(\omega t) \tag{1}$$

Where σ is the stress;
ε_o is the strain;
ω is the angular frequency;
t is the time;
E' is the *storage modulus* $\{E' = (\sigma_o/\varepsilon_o)\cos\delta\}$ and is a measure of recoverable strain energy in a deformed body;
E" is the *loss modulus* $\{E'' = (\sigma_o/\varepsilon_o)\sin\delta\}$ and is associated with the dissipation of energy as heat due to the deformation of the material;
δ is the phase angle.

The ratio E"/E' yields the loss tangent or damping factor (tanδ) which is the ratio of energy lost per cycle to the maximum energy stored and therefore recovered, per cycle. A typical dynamic mechanical analysis curve shows either E', E" or tanδ plotted as a function of time or temperature. In general, the most intense peak observed for either E" or tanδ in conjunction with a relatively pronounced drop in E' corresponds to the glass-transition temperature.

The transition temperature determined by dynamic techniques is both heating-rate dependent and frequency dependent. Therefore, the test method, heating rate, frequency, and the mechanical/rheological property (E', E", G', G" or tanδ) used to determine the T_g must be specified. E" has been preferentially used for T_g. In this regard, it has been found that the E" peak maximum at 1 Hz corresponded closely with the T_g obtained from volume-temperature measurements obtained by TMA [*21*].

As mentioned earlier, the change in the T_g (ΔT_g) as a tool to predict the service life of roof membranes is of critical importance. The ΔT_g can be used for establishing correlation between artificially aged (e.g., ultraviolet and heat) and naturally aged samples. T_g is an important indicator of the low temperature field service of a roof membrane. Below the T_g the material is more rigid, thus more prone to cracking, while above the T_g the material is more flexible, thus more resistant to cracking. It is important to note that both the ΔT_g as well as the absolute value of the T_g may be relevant in assessing the field service of a roof membrane. The ΔT_g would be more relevant as the absolute T_g approaches the in-service temperature of the membrane. Other factors such as, method of attachment (e.g., ballasted versus mechanically attached) and average environmental conditions (e.g., Texas versus Ontario) may influence the effectiveness of T_g as a predictive tool.

To this end, samples of a poly(vinyl chloride) (PVC) roof membrane based on a formulation with over 15 years of acceptable performance were either heated in a forced air oven or exposed to ultraviolet (UV) light in an artificial weathering device. In addition, samples of the same membrane, which have been exposed from 7 to 10 years in the Southern Ontario region were obtained for correlation with artificially aged samples.

Experimental Method

1. Exposure Conditions

Forty-one PVC roof samples were analyzed. The size of each of the samples was approximately 8 cm long and 5 cm wide (3 in. x 2 in.). The samples were exposed and labeled as shown (Table 1). The exposure conditions were as follows:

Heat Aged Samples - Specimens (3 in. x 2 in.) were die cut and placed on the center rack of a forced air oven at 80 °C or 100 °C (±2 °C). A set of two specimens were removed from the oven at 80 °C after 7, 14, 21, 28, 35, 42, 49 an 56 days while samples heated at 100 °C were removed after 7, 14, 21, 28 and 35 days. All specimens were rotated twice a week to ensure consistent thermal history. The specimens were removed from the oven and wrapped in aluminum foil.

Fluorescent UV/Condensation Exposure - Specimens were exposed as per ASTM Practice for Operating Light-and Water-Exposure Apparatus (Fluorescent UV/Condensation Type) for Exposure of Nonmetallic Materials G53, using UVB with peak emission at 313 nm, 8 hours UV and 4 hours condensation cycle at 50 °C. Specimens were removed after 1000 h, 2000 h, 3000 h, 4000 h and 5000 h.

Field Exposure - Four samples were exposed in the Southern Ontario region for 7, 9, and 10 years. This area is subject to wide swings in temperature, experiencing hot summers and cold winters.

2. Analysis

A Rheometric Scientific Solid Analyzer RSAII[TM3] equipped with an environmental controller was used to measure the glass transition temperature of the exposed and unexposed specimens. The calibration of the instrument was checked weekly and re-calibrated when required, using the manufacturer specification and procedures.

Samples were analyzed "as received". The glass transition temperature was measured according to ASTM D 6382 and ASTM D 5418. Rectangular strips were cut from the left and right side of each specimen, and placed on a dual cantilever fixture. If the T_g of the two specimens changed by more than 2 °C a third specimen was cut from the center. Each specimen was cooled to –80 °C. The temperature was allowed to stabilize for 5 minutes and increased to 50 °C at 2 °C/min. The test setup is given below:

Test type:	Dynamic temperature ramp
Sample Dimensions:	Length: fixed at 36.66 mm.
Frequency:	1 Hz (6.28 rad/s)
Temperature program:	-80 °C to 50 °C
Ramp rate:	2 °C/min

[3] Rheometric Scientific Inc., Piscataway, New Jersey.

Soak time:	1 minute
Time measurement:	1 minute
Strain:	0.01-0.02% depending on the specimen
AutoStrain:	On; adjustment 30% of current strain

TABLE 1 - *Sample labeling.*

No.	**Sample ID**	**Exposure Temp. (°C)**	**Exposure Time**		
			Hours	Days	Years
	Control 10449				
1	Specimen # 1	Control		0	
2	Specimen # 2	Control		0	
3	Specimen # 1	80 °C		7	
4	Specimen # 2	80 °C		7	
5	Specimen # 1	80 °C		14	
6	Specimen # 2	80 °C		14	
7	Specimen # 1	80 °C		21	
8	Specimen # 2	80 °C		21	
9	Specimen # 1	80°C		28	
10	Specimen # 2	80 °C		28	
11	Specimen # 1	80 °C		35	
12	Specimen # 2	80 °C		35	
13	Specimen # 1	80 °C		42	
14	Specimen # 2	80 °C		42	
15	Specimen # 1	80 °C		49	
16	Specimen # 2	80 °C		49	
17	Specimen # 1	80 °C		56	
18	Specimen # 2	80 °C		56	
19	Specimen # 1	100°C		7	
20	Specimen # 2	100 °C		7	
21	Specimen # 1	100 °C		14	
22	Specimen # 2	100 °C		14	
23	Specimen # 1	100°C		21	
24	Specimen # 2	100 °C		21	
25	Specimen # 1	100 °C		28	
26	Specimen # 2	100 °C		28	
27	Specimen # 1	100°C		35	
28	Specimen # 2	100 °C			
29	Sample # 1	Field (Southern Ontario)			9
30	Sample # 2	Field (Southern Ontario)			9
31	Sample # 3	Field (Southern Ontario)			7
32	Sample # 4	Field (Southern Ontario)			10
33	Specimen # 1	UV	1000		
34	Specimen # 2	UV	1000		
35	Specimen # 1	UV	2000		
36	Specimen # 2	UV	2000		
37	Specimen # 1	UV	3000		
38	Specimen # 2	UV	3000		
39	Specimen # 1	UV	4000		
40	Specimen # 2	UV	4000		
41	Specimen # 1	UV	5000		

The Tg was obtained from both, the intercept of two tangents on the storage modulus curve (E'_{int}) and the maximum of the loss modulus curve (E''_{max}). The fast Fourier transform (FFT) smoothing (4 points filter) routine from the RSI Orchestrator™[4] software was used to smooth the curves before plotting. All samples were run at least in duplicate.

Results and Discussion

The Tgs were obtained from both the E'_{int} and E''_{max} curves for the unexposed and exposed specimens. Figure 1 illustrates a typical DMA plot on which is noted the Tg as obtained from both the E'_{int} and E''_{max} curves. Table 2 (A-E) summarizes the average Tg and the ΔT_g of the samples analyzed. Dimensions and individual measurements for Tg as well as the E" curves of the samples were not included due to space considerations.

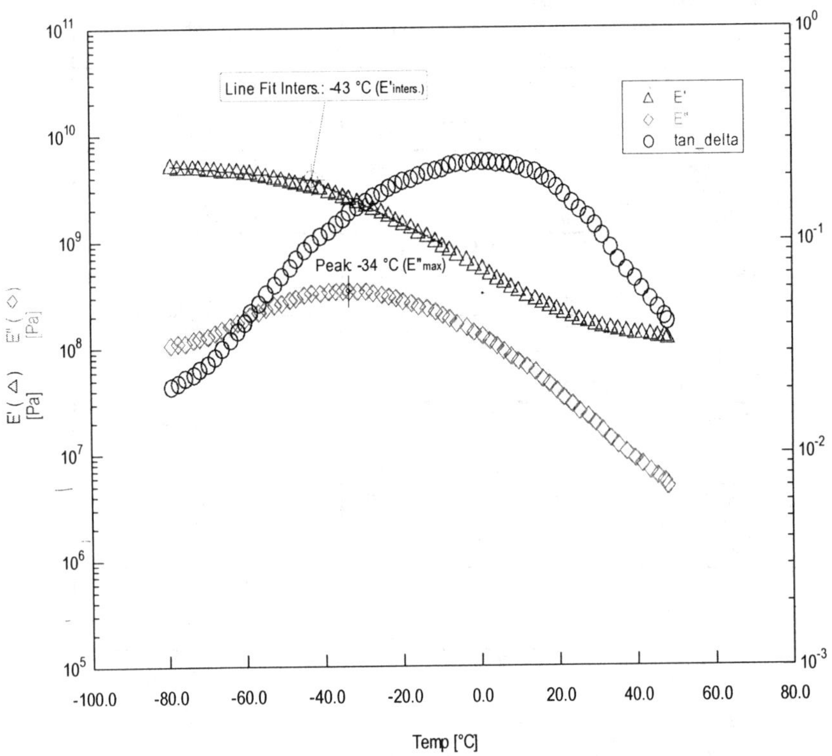

FIG. 1 - *DMA plot showing how T_g is obtained from E' and E" curves.*

[4] Rheometric Scientific Inc., Piscataway, New Jersey.

Table 2A summarizes the average T_g and the ΔT_g of the unexposed sample and those heated at 80 °C for up to 56 days. The T_g of the unexposed sample obtained from the E'_{int} and E''_{max} curves is –42 °C and –37 °C, respectively, whereas that of the heated samples ranges from –46 °C to –42 °C for the E'_{int} and from –38 °C to –35 °C for the E''_{max}.

Usually the T_g values from the E'_{int} and E''_{max} curves follow a similar trend. For example, an increase in the T_g from E'_{int} is followed by an increase in the T_g from the E''_{max}. However, that trend is not as clearly observed in the T_g of samples heated at 80 °C and 100 °C as well as the field and UV-exposed up to 5000 hours as shown in Tables 2A-2D.

TABLE 2A - *Average T_g and ΔTg of roof samples.*

Sample ID	Heat-Treatment	T_g^{**} (°C)			
		(E'_{int})	$\Delta T_g^{\S}$	(E''_{max})	$\Delta T_g^{\S}$
Control # 10499					
Specimen # 1	*Control*	–42±1		–36±1	
Specimen # 2	"	–41±1		–38±0	
Average	***Control***	***–42±0***	-	***–37±1***	-
Specimen # 1	*7 days at 80 °C*	–44±1		–39±0	
Specimen # 2	"	–41±4		–36±1	
Average	***7 days at 80 °C***	***–42±2***	0	***–38±2***	–1
Specimen # 1	14 days at 80 °C	–45±1		–37±3	
Specimen # 2	"	–45±1		–36±1	
Average	***14 days at 80 °C***	***–45±0***	–3	***–36±1***	+1
Specimen # 1	21 days at 80 °C	–45±1		–38±2	
Specimen # 2	"	–46±1		–39±0	
Average	***21 days at 80 °C***	***–46±1***	–4	***–38±1***	–1
Specimen # 1	28 days at 80 °C	–44±1		–38±1	
Specimen # 2	"	–45±1		–38±1	
Average	***28 days at 80 °C***	***–44±1***	–2	***–38±0***	–1
Specimen # 1	35 days at 80 °C	–46±1		–38±1	
Specimen # 2	"	–43±1		–36±2	
Average	***35 days at 80 °C***	***–44±2***	–2	***–37±1***	+0
Specimen # 1	42 days at 80 °C	–46±0		–36±1	
Specimen # 2	"	–44±0		–39±0	
Average	***42 days at 80 °C***	***–45±1***	–3	***–38±2***	–1
Specimen # 1	49 days at 80 °C	–44±0		–37±0	
Specimen # 2	"	–43±0		–36±1	
Average	***49 days at 80 °C***	***–44±1***	–2	***–36±1***	+1
Specimen # 1	56 days at 80 °C	–42±2		–34±1	
Specimen # 2	"	–42±1		–36±2	
Average	***56 days at 80 °C***	***–42±0***	0	***–35±1***	+2

$^{**}T_g$ is the average of T_g^{*} of specimens # 1&2

$^{\S}\Delta T_g = T_g^{**}$(exposed) – T_g^{**}(unexposed)

It is interesting to note that the T_g (E'_{int}) of the exposed samples is slightly cooler than that of the unexposed (negative ΔT_g) whereas the T_g (E''_{max}) of the same samples is, in general, slightly warmer (positive value). In general, a negative ΔT_g suggests breakdown of the polymer chain. If this is the case, then both ΔT_g should have negative values. This may result from experimental error due to the fact that the T_g (E'_{int}) is obtained from the intercept of two lines drawn manually on the E' curve. In spite of these differences, only the T_g values obtained from the E" curve will be discussed here.

Results from the E" curves suggest that heating at 80 °C up to 56 days did not affect the T_g of the specimens appreciably. The ΔT_g values (with an experimental error of ±2 °C) are well within the 8 °C increase limit recommended by the CIB/RILEM Committee [*2*].

The glass transition temperature (T_g) of specimens heated at 100 °C for 21 days (Table 2B) shows a similar trend to those heated at 80 °C. However, specimens heated for 28 and 35 days showed ΔT_gs (Table 2B) of 5 °C and 9 °C, respectively. A ΔT_g of 5 °C is below the 8 °C limit, while 9 °C is just outside the limit, recommended by the CIB/RILEM Committee [*2*]. Such an increase in T_g suggests some changes in the chemical composition of the material.

TABLE 2B - *Glass transition temperature (T_g^* in Table 2A) and ΔTg of roof samples.*

Sample ID	Heat- Treatment	T_g^{**} (°C)			
		(E'_{int})	$\Delta T_g^{§}$	(E''_{max})	$\Delta T_g^{§}$
Control # 10499					
Specimen # 1	*Control*	−42±1		−36±1	
Specimen # 2	"	−41±1		−38±0	
Average	***Control***	***−42±0***	-	***−37±1***	-
Specimen # 1	*7 days at 100 °C*	−48±1		−38±1	
Specimen # 2	"	−46±1		−37±1	
Average	***7 days at 100 °C***	***−47±1***	−6	***−38±1***	−1
Specimen # 1	*14 days at 100 °C*	−44±1		−36±1	
Specimen # 2	"	−44±1		−34±1	
Average	***14 days at 100 °C***	***−44±0***	−2	***−35±1***	+2
Specimen # 1	*21 days at 100 °C*	−44±0		−34±1	
Specimen # 2	"	−42±1		−34±1	
Average	***21 days at 100 °C***	***−43±1***	−1	***−34±0***	+3
Specimen # 1	*28 days at 100 °C*	−40±1		−31±2	
Specimen # 2	"	−39±1		−31±1	
Average	***28 days at 100 °C***	***−39±1***	+3	***−32±1***	+5
Specimen # 1	*35 days at 100 °C*	−38±1		−28±1	
Specimen # 2	"	−38±0		−29±0	
Average	***35 days at 100 °C***	***−38±0***	+4	***−28±1***	+9

$^{**}T_g$ is the average of T_g^* of specimens # 1&2
$^{§}\Delta T_g = T_g^{**}$(exposed) – T_g^{**}(unexposed)

There were no appreciable changes in the T_g (Table 2C) of the specimens exposed to UV up to 3000 hours. However, specimens exposed for 4000 hours show a ΔT_g of 3 °C whereas that exposed for 5000 hours has a ΔT_g of 7 °C.

TABLE 2C - *Summary of average T_g of UV-exposed samples at 50 °C.*

Sample ID	UV-Treatment	T_g^* °C			
		E'_{int}	$\Delta T_g^§$	E''_{max}	$\Delta T_g^§$
Control 10449	***Control***	−42±0	–	−37±1	–
Specimen # 1	***1000 hours***	−46±0	–	−39±0	–
Specimen # 2	"	−48±1	–	−39±0	-
Average	***1000 hours***	***−47±1***	−5	***-39±0***	−2
Specimen # 1	***2000 hours***	−44±1	–	−35±0	-
Specimen # 2	"	−43±1	–	−35±2	-
Average	***2000 hours***	***−44±1***	−2	***−35±0***	+2
Specimen # 1	***3000 hours***	−44±2	–	−35±2	-
Specimen # 2	"	−44±1	–	−35±0	-
Average	***3000 hours***	***−44±0***	−2	***−35±0***	+2
Specimen # 1	***4000 hours***	−42±0	–	−33±0	-
Specimen # 2	"	−42±1	–	−35±0	-
Average	***4000 hours***	***−42±0***	−2	***−34±1***	+3
Specimen # 1	***5000 hours***	−38±3	+4	−30±1	+7

$^*T_g$ is the average of 2-3 individual measurements
$^§\Delta T_g = T_g^{**}$(exposed) – T_g^{**}(unexposed)

Table 2D summarizes the glass transition temperature for the field-exposed samples. The four samples were exposed in the Southern Ontario region. Samples #1 and #2, which had been exposed in the field for 9 years, have a similar T_g value (−40 °C). Although, Sample #3 has less exposure time (7 years) than Sample #4 (10 years), their T_g value is approximately the same (Table 2D). Based on manufacturer-supplied data, it was determined that Sample *10449* could be used as a "control". Using this approach, it is important to notice that the ΔT_g of Field Samples #1 and #2 is about −4 °C and that of Field Samples #3 and #4 is about 4 °C. The T_g values for Field Samples #3 and #4 are very similar to those of the samples heated for 56 days at 80 °C, 14-21 days at 100 °C and exposed to UV for 2000-4000 hours. The negative ΔT_g values of Samples #1 and #2 once again raises the question of another mechanism in the aging process.

TABLE 2D - *Summary of average T_g of field-exposed samples.*

Sample ID	T_g^* °C			
	E'_{int}	$\Delta T_g^§$	E''_{max}	$\Delta T_g^§$
Control 10449	−42±0	–	−37±1	–
Field Sample # 1 (9 years)	−46±1	−4	−41±0	−4
Field Sample # 2 (9 years)	−46±1	−4	−40±1	−3
Field Sample # 3 (7years)	−42±0	0	−33±0	+4
Field Sample # 4 (10 years)	−44±1	−2	−34±1	+3

$^*T_g$ is the average of 2-3 individual measurements
$^§\Delta T_g = T_g^{**}$(exposed) – T_g^{**}(unexposed)

Conclusions

The results of the DMA analysis have shown that heating of specimens at 80 °C for up to 56 days and 100 °C for up to 21 days does not appreciably increase the T_g of the samples. Similarly, exposure to UV up to 4000 hours does not seem to have a great effect on the T_g of the samples. However, an appreciable increase in the T_g of the specimens was observed when they were exposed to UV for 5000 hours and also when heated at 100 °C for 35 days. Therefore, this is important to understand for establishing correlation between artificially and naturally aged specimens.

A comparison between the T_g of samples exposed in the Southern Ontario region for 7, 9 and 10 years, and that of the control used for the other samples indicates that the samples did not undergo substantial degradation due to exposure. The T_g values of Field Samples #3 and #4 are very similar to those of the samples heated for 56 days at 80 °C, 14-21 days at 100 °C and those exposed to UV for 2000-4000 hours.

The ΔT_g values of Field Samples #1 and #2 do not correlate as closely with the artificially aged specimens. The negative ΔT_g values may be indicative of a different aging mechanism. In these cases, both the E' and the E" curves are showing negative values, which would tend to reduce the likelihood that this is simply due to error.

In light of the fact that these roof membranes are all still successfully functioning in the field, what can be gleaned? The artificial aging methods of heating for 56 days at 80 °C, 21 days at 100 °C, and UV exposure up to 4000 hours all seem to perform reasonably well as predictors of the T_g of at least some field-aged membranes up to a 10 year time horizon.

Based on the findings in this study, T_g, as determined by DMA, in combination with artificial aging are potentially useful tools in predicting the low temperature field service of PVC roof membranes over time. It is worth noting that heat aging at 80 °C for 56 days, 100 °C for 21 days, or UV exposure at 4000 hours seems to be reasonable since the ΔT_g (E") of the aged samples were close to what was seen in the field. ASTM Specification for Poly(Vinyl Chloride) Sheet Roofing (D 4434) should consider incorporating T_g (E"), ΔT_g (E") in its next revision. Perhaps, a staged approach should be used. The first stage should specify heat aging at 80 °C for 56 days or 100 °C for 21 days and simply reporting the T_g (E") and ΔT_g (E"). Once sufficient field data has been acquired from a broader spectrum of the North American environments and correlated with heat-aged samples for purposes of validation, the establishment of values for T_g (E") and ΔT_g (E") may be possible.

References

[1] Farlling, M.S., "New Laboratory Procedures to Evaluate the Durability of Roofing Membranes," Appendix D in "Performance Testing of Roofing Membrane Materials," Recommendations of the *Conseil International du Batiment pour la Recherche l'Etude et la Documentation (CIB) W.83 and Réunion Internationale des Laboratoires d'Essai et de Recherche sur les Matériaux et les Constructions (RILEM) 75-SLR Joint Committee on Elastomeric, Thermoplastic and Modified Bitumen Roofing*, RILEM, Paris, France, November 1988.

[2] "Thermal Analysis Testing of Roofing Membrane Materials," Final Report of the Thermal Analysis Task Group, RILEM 120-MRS/CIB W.83 Joint Committee on Membrane Roofing Systems, Available from W.J. Rossiter, Building Materials Division, National Institute of Standards and Technology, Gaithersburg, MD December 1995.

[3] Cash, C.G., "Thermal Evaluation of One-Ply Sheet Roofing", *Single Ply Roofing Technology: ASTM STP 790*, 1982, pp. 55-64.

[4] Backenstow, D. and Flüeler, P., "Thermal Analysis for Characterization", *Proceedings, 9th Conference on Roofing Technology*, National Roofing Contractors Association, Rosemont, IL, April 1987, pp. 54-68.

[5] Gaddy, G.D., Rossiter, Jr., W.J., and Eby, R.K., "The Application of Thermal Analysis to the Characterization of EPDM Roofing Membrane Materials", *Roofing Research and Standards Development: ASTM STP 1088,* 1990, pp. 37-52.

[6] Dutt, O., Paroli, R.M., Mailvaganam, N.P., and Turenne, R.G., "Glass-Transitions in Polymeric Roofing Membranes", *Proceedings, 1991 International Symposium on Roofing Technology,* F. Kocich, editor, 1991, pp. 495-501.

[7] Paroli, R.M. and Dutt, O., "Dynamic Mechanical Analysis Studies of Reinforced Polyvinyl Chloride (PVC) Roofing Membranes", *Polymeric Materials Science and Engineering,* Vol. 65, 1991, pp. 362-363.

[8] Paroli, R.M., Dutt, O., Delgado, A.H., and Mech, M., "The Characterization of EPDM Roofing Membranes by Thermogravimetry and Dynamic Mechanical Analysis", *Thermochimica Acta,* Vol. 182, 1991, pp. 303-317.

[9] Gaddy, G.D., Rossiter, Jr., W.J., and Eby, "The Application of Thermal Analysis to the Characterization of EPDM Roofing Membrane Materials After Exposure in Service", *Proceedings, 1991 International Symposium on Roofing Technology,* F. Kocich, editor, 1991, pp. 502-507.

[10] Paroli, R.M., Dutt, O., Delgado, A.H., and Stenman, H.K., "Ranking Polyvinyl Chloride (PVC) Roofing Membranes using Thermal Analysis", *Journal of Materials in Civil Engineering*, Vol. 5, No. 1, 1993, pp. 83-95.

[11] Gaddy, G.D., Rossiter, Jr., W.J., and Eby, R.K., "The Use of TMA to Characterize EPDM Roofing Membrane Materials", *ASTM STP 1136*, 1991, pp. 168-175.

[12] Penn, J.J. and Paroli, R.M., "Evaluating the Effects of Aging on the Thermal Properties of EPDM Roofing Materials", *Proceedings of the 21st North American Thermal Analysis Society (NATAS) Conference Atlanta, Georgia,* Sept.13-18, 1992, pp. 612-617.

[13] Oba, K. and Björk, F., "Dynamic Mechanical Properties of Single-Ply Roof Coverings for Low-Slope Roofs and the Influence of the Water", *Polymer Testing*, Vol. 12, 1993, pp. 35-56.

[14] Oba, K. and Björk, F., "The Influence of Water on the Mechanical Strength of Seams in Single-Ply Coverings for Low-Slope Roofs", *Kautschuk und Gummi, Kunststoffe*, Vol. 45, No. 3, 1992, pp. 195-202.

[15] Paroli, R.M., Smith, T.L., and Whelan, B., "Shattering of Unreinforced PVC Roof Membranes: Problem Phenomenon, Causes and Prevention", *NRCA/NIST Tenth Conference on Roofing Technology,* April 22-23, 1993, pp. 93-107.

[16] Penn, J.J. and Paroli, R.M., "Evaluating the Effects of Aging on the Thermal Properties of EPDM Roofing Membranes", *Thermochimica Acta*, Vol. 226, 1993, pp. 77-84.

[17] Rodriguez, I., Dutt, O., Paroli, R.M., and Mailvaganam, N.P., "Effect of Heat-Ageing on the Thermal and Mechanical Properties of APP- and SBS-Modified Bituminous Roofing Membranes", *Materials and Structures,* Vol. 26, No. 160, 1993, pp. 355-361.

[18] Oba, K., "Flat Roofs Investigation of Heat Welding Techniques for Polymer-Modified Bituminous Roofing Membranes." Dissertation, Royal Institute of Technology, Sweden, 1994.

[19] Oba, K. and Partl, M.N., "Seam Performance of Polymer-Modified Bituminous Roofing Sheets", EMPA-Forschungs- und Arbeitsbericht No. 136/6: FE 147′135. Swiss Federal Laboratories for Materials Testing and Research, Switzerland, 1994.

[20] Oba, K. and Roller, A., "Characterization of Polymer Modified Bituminous Roofing Membranes Using Thermal Analysis", *Materials and Structures/Materiaux et Constructions,* Vol. 28, No. 184, 1995, pp. 596-603.

[21] Paroli, R.M. and Penn, J.J., "Measuring the Glass-Transition Temperature of EPDM Roofing Materials: Comparison of DMA, TMA and DSC Techniques", *Assignment of the Glass Transition, ASTM STP 1249,* R.J. Seyler, Ed., American Society for Testing and Materials, Philadelphia, 1994, pp. 269-276.

[22] Paroli, R.M., Dutt, O., Smith, T.L., and Whelan, B.J., "Enhancement of the Standard for PVC Roof Membrane Material with the use of Mechanical and Chemical Analysis", *Roofing Research and Standards Development: Volume 3, ASTM STP 1224*, T.J. Wallace and W.J. Rossiter, Jr., Eds., American Society for Testing and Materials, Philadelphia, 1994, pp. 139-147.

[23] Paroli, R.M., Dutt, O., Delgado, A.H., and Lei, W., "A Novel Approach to Investigating the Durability of Roofing Membranes", *Polymeric Materials Science and Engineering*, Vol. 72, 1995, pp. 378-379.

Ken Chernotowich,[1] Kenneth J. Brzozowski,[2] and Paul J. Sanicky[3]

Effects of Application Technique and Temperature on Characteristics of Modified Bitumen Membrane Assemblies

---

Reference: Chernotowich, K., Brzozowski, K. J., and Sanicky, P. J., **"Effects of Application Technique and Temperature on Characteristics of Modified Bitumen Membrane Assemblies,"** *Roofing Research and Standards Development: 5th Volume, ASTM STP 1451,* W. J. Rossiter and T. J. Wallace, Eds., ASTM International, West Conshohocken, PA, 2003.

Abstract: Results are presented and discussed of a laboratory study that measures T-peel interply adhesion of polymer modified bituminous membranes assembled using typical field application techniques and a wide range of asphalt temperatures. This evaluation includes both standard and polymer modified asphalts, and additionally examines factors such as: a) interply thickness, b) tendencies for the inclusion of blisters, and c) the effects of dark oven heat aging at 70°C (158°F) of the completed assemblies. The findings indicate that the use of elevated asphalt application temperatures, conventionally proposed to improve adhesion, has little to no effect on interply T-peel strengths. The results also illustrate patterns of interply thickness as it relates to asphalt application temperature and/or application technique. Additional indications reveal a tendency to develop blisters that are dependent on application technique and asphalt temperature. The study also proposes that if interply adhesion is considered to be a primary criterion for improved performance of polymer modified bituminous sheet systems, the use of polymer modified interply asphalts must be given due consideration for use in this application.

Keywords: asphalt, polymer modified asphalt, heat aging, membrane assembly temperature, membrane assembly technique, polymer modified bituminous membranes, T-peel adhesion

---

[1] 1976 Trailwood Drive, Twinsburg, OH 44087-2815.

[2] Director, Research and Development, W. P. Hickman Systems, Inc., 30700 Solon Industrial Parkway, Solon, OH 44139.

[3] Laboratory Technician, W. P. Hickman Systems, Inc., 30700 Solon Industrial Parkway, Solon, OH 44139.

Introduction

Over the past five years or so, the roofing community has experienced efforts to eliminate the use of equiviscous temperature (EVT) for asphalt application as it pertains to the construction of polymer modified bitumen membrane systems. To this end, the American Society for Testing and Materials (ASTM) modified the EVT definition in Standard Terminology Relating to Roofing, Waterproofing, and Bituminous Materials (D 1079) to indicate that EVT may not be appropriate for use with polymer modified bitumen products. Simply stated, EVT represents the temperature at which asphalt reaches a specified viscosity established to achieve a desired rate of interply quantity during application.

The main premise of this movement is to allow the use of higher asphalt application temperatures with the goal of inducing better fusion between the mopping asphalt and SBS modified sheets, thus increasing adhesion and minimizing blistering within a polymer modified membrane-roofing system. There have been several trade journal articles and position papers addressing the phenomenon of blistering, potential causes, and the actions or practices necessary to make positive corrections. A few articles that address these and attendant issues are references [*1-5*]. Also, reference [*6*] takes a look at the various field application techniques, and the variables each presents.

Yet despite the ASTM definition change, there seems to be a bit of confusion within the industry. As reference [*4*] points out, most manufacturers of SBS modified asphalt sheets continue to recommend application temperatures for the mopping asphalts that are either the EVT, or are within the accepted norm of ± 14 °C (25 °F) EVT range. Some manufacturers are more specific in that the specified temperature must be at "point of contact." In total, these recommendations clearly do not support the suggestion for higher application temperatures, and there is no indication as to exactly what is considered to be "acceptably higher."

This paper is the authors' second study that addresses whether or not increasing application temperatures will in fact provide increased adhesion and result in less blistering. The first study (reference [7]) included a wide range of application temperatures, evaluations with a polyester-reinforced standard base sheet, both standard and SEBS modified asphalts, preheating of sheets, and the removal of excess mineral partitioning agent. The current study concentrates on variable application temperatures and techniques, as well as a variety of interply asphalts.

Experimental Details

Raw Materials and Sampling

Each product used in the preparation of the polymer modified bitumen assemblies was randomly obtained from stock or purchased from a distributor. Each material was tested and found to conform to their particular specification and type as stated in the respective ASTM standard, the identities of which are as follows.

1) ASTM Specification for Asphalt Used in Roofing (D 312), Types III and IV.
2) ASTM Specification for SEBS Modified Mopping Asphalt Used in Roofing (D 6152).
3) ASTM Specification for Styrene Butadiene Styrene (SBS) Modified Bituminous Sheet Material Using a Polyester Reinforcement (D 6164), Type I, Grade S and Type I, Grade G.

The sheet reinforcements were randomly cut into 305 x 305 mm (12 x 12 in.) specimens, avoiding usage of the outer 51 mm (2 in.) along both edges. The respective stacks of cut specimens were then shuffled to further randomize their selection throughout the assembly process. It was at this point that thickness measurements for the two SBS modified sheets were determined.

As for the various mopping asphalts used throughout this study, sampling was done by randomizing collected pieces into a group of gallon containers so as to avoid any additional heat history prior to actual assembly of the test specimens. Within any one series, the mopping asphalt was heated to the lowest temperature, and then sequenced to the final temperature for the assemblies.

Test Design

A complete membrane set was designed to include: 1) each of the three standard application techniques, i.e., roll-in (setting and rolling), flying-in, and mop-and-flop; 2) three mopping asphalts; and 3) a range of four application temperatures to include 218°C (425°F), 232°C (450°F), 246°C (475°F), and 260°C (500°F). These temperatures essentially begin with an accepted common EVT for D 312, Type III asphalt, progressing through 260°C. The latter temperature allows the authors to maintain the accepted 14°C (25°F) safety factor below potential asphalt flash points. Eventually four types of mopping asphalts were selected, including D 312 Types III and IV, and two D 6152 SEBS modified asphalts differentiated by manufacturing process.

Membrane and Specimen Preparation

Assembling the 305 x 305 mm sample membranes for adhesion testing followed the guidelines of ASTM Test Method for Peel Resistance of Adhesives (T-Peel Test) (D 1876). Tape was applied 76 mm (3 in.) along one edge of both the lower (3mm smooth surface) and upper (4 mm granulated) sheets. Asphalt was heated, poured, and brushed, then the top reinforcement put in place according to the application technique being prepared. Each assembly was then immediately rolled with a rolling pin weighing approximately 2080 g (4.59 lb) to ensure proper embedment and consistency. The completed membranes were then allowed to cool overnight.

For preparation of the individual test specimens, each 305 x 305 mm sample membrane was cut using a 305 x 305 mm die specially designed to cut one half of the membrane into six 25.4 mm by 305 mm (1 x 12 in.) strip specimens, and the other half kept as an intact piece. The outside 25.4 x 305 mm edge strip was discarded to minimize any potential edge effects. The one-half piece was then subjected to 30 days dark oven aging at 70°C (158°F). Once the samples were removed from the oven they were permitted to cool to room temperature overnight, then sectioned into five strip specimens, positioning the die to remove the outer 12.7 mm (0.5 in.) edge from both sides, again addressing edge effects.

Test Procedures

All testing was performed on an Instron Model 4465 with fully integrated Series IX software for data collection. All ASTM D 1876 T-peel adhesion testing was performed at the ambient temperature of 23°C (73°F). Pneumatic grips were used with a jaw separation of 76 mm (3 in.) and crosshead speed of 51 mm/min (2 in./min). A full 178 mm (7 in.) of the bond line were tested, with data collection beginning after the first 25 mm (1 in.) of separation and terminating after an additional 140 mm (5.5 in.) of separation. Jaw separation continued for another 12.7 mm (0.5 in.) before the test was terminated, ensuring a continue load application throughout the test area.

Results and Discussion

For the first complete set, a standard ASTM D 312, Type III and two ASTM D 6152 SEBS modified mopping asphalts were chosen. The latter two represent products that were manufactured using either low (L) or high (H) shear mixing. Table 1 presents the data for the first set of 25.4 mm x 305 mm specimens aged one day under ambient conditions.

Examining each column by itself, the indication is that increasing application temperatures fail to improve adhesion, whether looking at peak or average peel loads, and regardless of asphalt type. Evaluating any asphalt across the three application procedures, there again appears to be very little difference. Results with the two SEBS modified asphalts are also equivalent. There are some data anomalies, such as the specimens for T. III at 246°C (fly-in) and SEBS (H) at 218°C (fly-in). Both of these groups exhibited failures exclusively off the bottom side of the 4 mm cap. There were blisters present with a clear transfer of partitioning agent into the asphalt.

In a very generic review, the location for adhesion separation was 67% from the top of the 3 mm base ply, 22% from the bottom of the cap ply, and 11% representing either a split between the two surfaces or cohesive failure of the asphalt. A variant that is indicated with the presence of voids is when they tended to occur. Chart A below

Table 1 - *T-Peel Adhesion for MB Assemblies*[a] *Aged 24 hrs. @ RT*

	Roll-In			Fly-In			Mop & Flop		
	T. III	SEBS(H)	SEBS(L)	T. III	SEBS(H)	SEBS(L)	T. III	SEBS(H)	SEBS(L)
Application Temp., 218°C (425°F)									
Max. Peel Load, N (lbf)	44.9 (10.1)	53.8 (12.1)	78.8 (17.7)	49.4 (11.1)	61.9 (13.9)	92.1 (20.7)	49.8 (11.2)	95.7 (21.5)	82.8 (18.6)
Std. Dev.	4.16 (0.94)	4.0 (0.91)	2.99 (0.67)	5.07 (1.14)	13.8 (3.09)	28.8 (6.46)	4.67 (1.05)	7.16 (1.61)	15.2 (3.42)
Avg. Peel Load, N (lbf)	39.2 (8.8)	47.6 (10.7)	60.5 (13.6)	43.6 (9.8)	29.8 (6.7)	68.5 (15.4)	42.3 (9.5)	57.9 (13.0)	71.2 (16.0)
Std. Dev.	2.59 (0.58)	3.52 (0.79)	1.98 (0.44)	5.61 (1.26)	3.10 (0.70)	11.5 (2.56)	3.83 (0.86)	8.23 (1.85)	14.2 (3.20)
Application Temp., 232°C (450°F)									
Max. Peel Load, N (lbf)	49.4 (11.1)	67.2 (15.1)	65.9 (14.8)	39.6 (8.9)	73.0 (16.4)	85.9 (19.3)	50.7 (11.4)	85.0 (19.1)	84.6 (19.0)
Std. Dev.	3.61 (0.81)	3.07 (0.69)	2.93 (0.66)	1.30 (0.29)	13.1 (2.93)	13.3 (2.98)	6.23 (1.40)	22.4 (5.03)	7.43 (1.67)
Avg. Peel Load, N (lbf)	41.8 (9.4)	53.8 (12.1)	53.8 (12.1)	33.4 (7.5)	49.4 (11.1)	59.6 (13.4)	40.9 (9.2)	57.0 (12.8)	66.3 (14.9)
Std. Dev.	4.86 (1.09)	2.59 (0.58)	0.96 (0.22)	1.14 (0.26)	12.4 (2.80)	7.69 (1.73)	3.68 (0.83)	6.02 (1.35)	8.93 (2.01)
Application Temp., 246°C (475°F)									
Max. Peel Load, N (lbf)	49.8 (11.2)	65.9 (14.8)	80.1 (18.0)	35.6 (8.0)	64.5 (14.5)	83.7 (18.8)	52.1 (11.7)	85.9 (19.3)	120.2 (27.0)
Std. Dev.	10.9 (2.44)	6.18 (1.39)	10.4 (2.33)	1.50 (0.39)	21.2 (4.77)	7.84 (1.76)	3.59 (0.81)	11.3 (2.55)	9.40 (2.11)
Avg. Peel Load, N (lbf)	41.8 (9.4)	58.7 (13.2)	54.3 (12.2)	24.5 (5.5)	48.5 (10.9)	64.1 (14.4)	44.5 (10.0)	55.6 (12.5)	86.8 (19.5)
Std. Dev.	14.3 (3.22)	3.74 (0.84)	2.39 (0.54)	4.17 (0.94)	7.16 (1.61)	7.82 (1.76)	5.08 (1.14)	2.33 (0.52)	12.0 (2.69)
Application Temp., 260°C (500°F)									
Max. Peel Load, N (lbf)	48.1 (10.8)	63.6 (14.3)	71.6 (16.1)	45.8 (10.3)	70.8 (15.9)	77.4 (17.4)	51.6 (11.6)	103.2 (23.2)	79.7 (17.9)
Std. Dev.	1.50 (0.38)	1.50 (0.34)	9.23 (2.08)	7.57 (1.70)	3.85 (0.87)	12.6 (2.82)	1.78 (0.4)	4.98 (1.12)	8.48 (1.91)
Avg. Peel Load, N (lbf)	43.2 (9.7)	56.5 (12.7)	52.5 (11.8)	38.7 (8.7)	42.7 (9.6)	53.4 (12.0)	42.7 (9.6)	56.1 (12.6)	49.4 (11.1)
Std. Dev.	0.75 (0.17)	1.76 (0.40)	1.22 (0.28)	6.78 (1.52)	4.89 (1.10)	3.58 (0.81)	4.20 (0.94)	5.44 (1.22)	4.52 (1.02)

Note

[a]Modified bitumen assemblies consist of one ply 3 mm smooth-surfaced base and one ply 4 mm granulated cap.

illustrates each group and whether or not any voids were present. No attempt was made to quantify the number or qualify by void size. Rather, if only one test strip was found to contain even a small void, it was given a "V" in Chart A. Obviously there were different sizes and quantities from test group to test group, specimen to specimen. This is merely a presentation of tendencies from one application procedure to another, in this laboratory analysis. The authors make no judgment as to any implications of what may occur during field assembly. However, under the tightly controlled conditions of this laboratory study, these results indicate that the application techniques of fly-in and mop-and-flop exhibit tendencies for increased levels of voids.

Chart A

	Roll-In			**Fly-In**			**Mop-and-Flop**		
	T. III	**H**	**L**	**T. III**	**H**	**L**	**T. III**	**H**	**L**
218 °C	V	OK	OK	V	V	V	V	OK	OK
232 °C	OK	OK	OK	OK	V	V	OK	OK	V
246 °C	V	OK	OK	V	V	V	V	OK	V
260 °C	OK	OK	OK	V	V	V	V	OK	OK

The data in Table 2 illustrate the results after 30 days aging at 70°C. Once again if we look at each asphalt column, increasing application temperature provides no improvement, and across the three application procedures also shows equivalent performance. As for the two SEBS modified asphalts, the low shear version demonstrated significantly less loss of adhesion after the aging process.

For the second complete series of samples presented in Tables 3 and 4, use of the ASTM D 312, Type III and ASTM D 6152 SEBS (L) asphalts were repeated, while ASTM D 312, Type IV was added. Again looking at the respective columns in both tables, neither increasing application temperature nor any of the application techniques demonstrated an improvement of adhesion. When comparing ASTM D 312 Types III and IV, the former proved to be the better performer, particularly with the aged samples. The latter demonstrated a consistent 50% decrease in adhesion in all three techniques. As for the SEBS specimens, there was a significant improvement in adhesion within each application technique.

Finally, Chart B below presents the measured thickness (mm) of the individual strip specimens just prior to T-peel testing. As expected, there is a general thinning of the interply mopping with increasing temperature. The results for the roll-in grouping were quite consistent. Some of the specimens had an interply layer that was quite thin. The combined average thickness with the two modified sheets equals

Table 2 - *T-Peel Adhesion for MB Assemblies*[a] *Aged 30 days @ 70° C (158° F)*

	Roll-In			Fly-In			Mop & Flop		
	T. III	SEBS(H)	SEBS(L)	T. III	SEBS(H)	SEBS(L)	T. III	SEBS(H)	SEBS(L)
Application Temp., 218°C (425°F)									
Max. Peel Load, N (lbf)	29.8 (6.7)	44.9 (10.1)	70.3 (15.8)	40.5 (9.1)	60.1 (13.5)	78.8 (17.7)	44.1 (9.9)	44.9 (10.1)	93.9 (21.1)
Std. Dev.	3.49 (0.79)	2.45 (0.55)	4.84 (1.09)	10.8 (2.43)	8.79 (1.98)	7.56 (1.70)	14.8 (3.33)	3.03 (0.68)	5.89 (1.32)
Avg. Peel Load, N (lbf)	24.5 (5.5)	36.0 (8.1)	42.7 (9.6)	28.0 (6.3)	37.4 (8.4)	62.7 (14.1)	27.6 (6.2)	34.3 (7.7)	73.9 (16.6)
Std. Dev.	2.83 (0.64)	0.67 (0.15)	4.61 (1.04)	2.93 (0.66)	1.42 (0.32)	7.84 (1.76)	2.35 (0.53)	0.68 (0.15)	5.44 (1.22)
Application Temp., 232°C (450°F)									
Max. Peel Load, N (lbf)	30.3 (6.8)	38.3 (8.6)	66.3 (14.9)	36.5 (8.2)	54.3 (12.2)	68.5 (15.4)	26.7 (6.0)	56.1 (12.6)	83.2 (18.7)
Std. Dev.	1.12 (0.25)	1.46 (0.33)	5.59 (1.26)	3.22 (0.72)	14.3 (3.21)	12.0 (2.69)	2.03 (0.46)	2.57 (0.58)	6.26 (1.41)
Avg. Peel Load, N (lbf)	24.9 (5.6)	32.5 (7.3)	43.2 (9.7)	26.7 (6.0)	35.2 (7.9)	54.3 (12.2)	21.4 (4.8)	37.8 (8.5)	68.5 (15.4)
Std. Dev.	0.45 (0.10)	0.72 (0.16)	2.54 (0.57)	1.41 (0.32)	1.02 (0.23)	10.5 (2.35)	2.63 (0.59)	2.67 (0.60)	5.66 (1.27)
Application Temp., 246°C (475°F)									
Max. Peel Load, N (lbf)	29.4 (6.6)	41.4 (9.3)	49.8 (11.2)	36.5 (8.2)	51.6 (11.6)	74.3 (16.7)	35.6 (8.0)	48.5 (10.9)	89.0 (20.0)
Std. Dev.	3.82 (0.86)	3.4 (0.77)	8.02 (1.80)	6.21 (1.40)	3.73 (0.84)	8.50 (1.91)	12.6 (2.82)	9.05 (2.03)	3.02 (0.68)
Avg. Peel Load, N (lbf)	24.5 (5.5)	34.3 (7.7)	30.3 (6.8)	22.3 (5.0)	33.8 (7.6)	55.2 (12.4)	26.7 (6.0)	32.9 (7.4)	72.5 (16.3)
Std. Dev.	3.84 (0.86)	0.83 (0.19)	3.04 (0.68)	4.40 (0.99)	1.04 (0.23)	7.14 (1.61)	4.90 (1.10)	1.82 (0.41)	6.52 (1.47)
Application Temp., 260°C (500°F)									
Max. Peel Load, N (lbf)	32.0 (7.2)	39.2 (8.8)	62.7 (14.1)	31.6 (7.1)	46.3 (10.4)	72.5 (16.3)	27.1 (6.1)	44.1 (9.9)	77.0 (17.3)
Std. Dev.	4.15 (0.93)	1.46 (0.33)	9.63 (2.17)	5.87 (1.32)	4.48 (1.01)	8.43 (1.90)	6.10 (1.37)	4.20 (0.94)	6.76 (1.52)
Avg. Peel Load, N (lbf)	24.0 (5.4)	32.0 (7.2)	30.7 (6.9)	24.0 (5.4)	34.7 (7.8)	45.4 (10.2)	20.5 (4.6)	30.7 (6.9)	55.2 (12.4)
Std. Dev.	0.6 (0.13)	0.85 (0.19)	6.36 (1.43)	1.74 (0.39)	1.13 (0.26)	4.07 (0.91)	3.70 (0.83)	1.17 (0.26)	3.14 (0.71)

Note

[a]Modified bitumen assemblies consist of one ply 3 mm smooth-surfaced base and one ply 4 mm granulated cap.

Table 3 - *T-Peel Adhesion for MB Assemblies*[a] *Aged 24 hrs. @ RT*

	Roll-In			Fly-In			Mop & Flop		
	T. III	T. IV	SEBS(L)	T. III	T. IV	SEBS(L)	T. III	T. IV	SEBS(L)
Application Temp., 218°C (425°F)									
Max. Peel Load, N (lbf)	33.4 (7.5)	21.4 (4.8)	84.6 (19.0)	28.5 (6.4)	19.6 (4.4)	89.0 (20.0)	29.4 (6.6)	26.7 (6.0)	133.5 (30.0)
Std. Dev.	3.21 (0.72)	1.44 (0.32)	11.9 (2.67)	4.92 (1.11)	1.85 (0.42)	24.1 (5.42)	1.18 (0.27)	5.98 (1.34)	6.96 (1.57)
Avg. Peel Load, N (lbf)	24.5 (5.5)	18.2 (4.1)	67.2 (15.1)	22.3 (5.0)	14.7 (3.3)	64.5 (14.5)	26.3 (5.9)	19.6 (4.4)	107.7 (24.2)
Std. Dev.	1.26 (0.28)	0.44 (0.10)	7.41 (1.67)	7.0 (1.58)	4.10 (0.92)	22.0 (4.95)	1.09 (0.25)	0.98 (0.22)	8.14 (1.83)
Application Temp., 232°C (450°F)									
Max. Peel Load, N (lbf)	26.7 (6.0)	22.7 (5.1)	105.9 (23.8)	31.2 (7.0)	27.6 (6.2)	120.2 (27.0)	40.1 (9.0)	22.3 (5.0)	153.1 (34.4)
Std. Dev.	2.88 (0.65)	2.80 (0.63)	6.6 (1.48)	2.88 (0.65)	5.14 (1.16)	19.3 (4.34)	7.92 (1.78)	4.12 (0.93)	16.1 (3.61)
Avg. Peel Load, N (lbf)	22.7 (5.1)	16.0 (3.6)	67.2 (15.1)	25.4 (5.7)	19.6 (4.4)	89.0 (20.0)	29.4 (6.6)	16.5 (3.7)	126.4 (28.4)
Std. Dev.	1.29 (0.29)	3.32 (0.75)	6.73 (1.51)	3.60 (0.81)	0.99 (0.22)	24.9 (5.60)	9.34 (2.10)	2.67 (0.60)	14.8 (3.33)
Application Temp., 246°C (475°F)									
Max. Peel Load, N (lbf)	29.4 (6.6)	22.7 (5.1)	85.9 (19.3)	30.7 (6.9)	26.7 (6.0)	105.9 (23.8)	27.6 (6.2)	24.9 (5.6)	126.8 (28.5)
Std. Dev.	2.81 (0.63)	2.46 (0.55)	8.44 (1.90)	6.79 (1.53)	1.66 (0.37)	15.2 (3.42)	1.61 (0.36)	2.0 (0.45)	3.67 (0.83)
Avg. Peel Load, N (lbf)	23.6 (5.3)	17.4 (3.9)	53.8 (12.1)	21.8 (4.9)	19.6 (4.4)	60.1 (13.5)	24.5 (5.5)	20.0 (4.5)	96.1 (21.6)
Std. Dev.	1.38 (0.31)	1.32 (0.30)	2.95 (0.66)	9.38 (2.11)	1.59 (0.36)	11.9 (2.68)	1.89 (0.43)	0.83 (0.19)	6.49 (1.46)
Application Temp., 260°C (500°F)									
Max. Peel Load, N (lbf)	33.8 (7.6)	29.8 (6.7)	86.8 (19.5)	27.1 (6.1)	26.7 (6.0)	84.6 (19.0)	28.5 (6.4)	20.9 (4.7)	139.3 (31.3)
Std. Dev.	3.58 (0.81)	6.51 (1.46)	20.4 (4.60)	2.96 (0.67)	2.64 (0.59)	8.14 (1.83)	7.70 (1.73)	2.32 (0.52)	15.1 (3.40)
Avg. Peel Load, N (lbf)	22.7 (5.1)	18.7 (4.2)	58.7 (13.2)	19.6 (4.4)	20.5 (4.6)	65.4 (14.7)	21.8 (4.9)	18.2 (4.1)	83.6 (18.8)
Std. Dev.	1.87 (0.42)	1.93 (0.43)	13.7 (3.07)	2.05 (0.46)	1.15 (0.26)	6.91 (1.55)	2.63 (0.59)	0.73 (0.17)	31.9 (7.18)

Note

[a]Modified bitumen assemblies consist of one ply 3 mm smooth-surfaced base and one ply 4 mm granulated cap.

Table 4 - *T-Peel Adhesion for MB Assemblies*[a] *Aged 30 days @ 70° C (158° F)*

	Roll-In			Fly-In			Mop & Flop		
	T. III	**T. IV**	**SEBS(L)**	**T. III**	**T. IV**	**SEBS(L)**	**T. III**	**T. IV**	**SEBS(L)**
Application Temp., 218°C (425°F)									
Max. Peel Load, N (lbf)	24.5 (5.5)	11.6 (2.6)	60.5 (13.6)	23.6 (5.3)	11.8 (2.6)	54.7 (12.3)	21.8 (4.9)	13.4 (3.0)	105.0 (23.6)
Std. Dev.	2.86 (0.64)	0.43 (0.10)	9.69 (2.18)	1.75 (0.39)	1.60 (0.36)	14.6 (3.29)	0.52 (0.12)	1.33 (0.30)	14.0 (3.14)
Avg. Peel Load, N (lbf)	21.8 (4.9)	8.6 (1.9)	43.2 (9.7)	20.5 (4.6)	8.5 (1.9)	29.4 (6.6)	20.0 (4.5)	9.8 (2.2)	79.7 (17.9)
Std. Dev.	2.49 (0.56)	0.71 (0.16)	4.73 (1.06)	0.32 (0.07)	0.43 (0.10)	3.89 (0.88)	0.51 (0.11)	1.65 (0.37)	6.21 (1.40)
Application Temp., 232°C (450°F)									
Max. Peel Load, N (lbf)	23.1 (5.2)	14.2 (3.2)	94.8 (21.3)	20.9 (4.7)	14.7 (3.3)	91.2 (20.5)	22.3 (5.0)	13.8 (3.1)	108.1 (24.3)
Std. Dev.	3.71 (0.83)	1.88 (0.42)	5.81 (1.31)	1.05 (0.24)	1.61 (0.36)	4.99 (1.12)	1.30 (0.29)	2.0 (0.45)	13.7 (3.08)
Avg. Peel Load, N (lbf)	17.8 (4.0)	9.3 (2.1)	71.2 (16.0)	18.2 (4.1)	10.7 (2.4)	68.5 (15.4)	19.6 (4.4)	10.2 (2.3)	82.3 (18.5)
Std. Dev.	0.67 (0.15)	0.59 (0.13)	5.40 (1.21)	0.52 (0.12)	0.89 (0.20)	9.49 (1.68)	0.95 (0.21)	1.06 (0.24)	10.2 (2.29)
Application Temp., 246°C (475°F)									
Max. Peel Load, N (lbf)	20.0 (4.5)	11.6 (2.6)	25.4 (5.7)	20.9 (4.7)	14.2 (3.2)	46.3 (10.4)	27.1 (6.1)	13.8 (3.1)	70.3 (15.8)
Std. Dev.	0.75 (0.17)	2.18 (0.49)	2.26 (0.51)	1.79 (0.40)	3.58 (0.81)	9.12 (2.05)	12.4 (2.78)	0.70 (0.16)	12.5 (2.80)
Avg. Peel Load, N (lbf)	17.4 (3.9)	8.0 (1.8)	13.8 (3.1)	18.7 (4.2)	8.5 (1.9)	25.4 (5.7)	18.7 (4.2)	10.7 (2.4)	44.5 (10.0)
Std. Dev.	0.47 (0.11)	0.36 (0.08)	1.89 (0.42)	0.66 (0.15)	0.49 (0.11)	3.59 (0.81)	0.54 (0.12)	21.0 (0.47)	11.2 (2.52)
Application Temp., 260°C (500°F)									
Max. Peel Load, N (lbf)	20.5 (4.6)	17.4 (3.9)	33.8 (7.6)	21.4 (4.8)	14.7 (3.3)	39.2 (8.8)	22.3 (5.0)	13.8 (3.1)	90.8 (20.4)
Std. Dev.	0.18 (0.04)	8.62 (1.94)	5.68 (1.28)	1.35 (0.30)	4.04 (0.91)	12.7 (2.86)	3.46 (0.78)	9.29 (2.09)	14.2 (3.18)
Avg. Peel Load, N (lbf)	18.2 (4.1)	8.9 (2.0)	19.1 (4.3)	19.6 (4.4)	9.8 (2.2)	21.8 (4.9)	19.1 (4.3)	8.0 (1.8)	78.8 (17.7)
Std. Dev.	0.24 (0.06)	0.42 (0.10)	2.94 (0.66)	1.02 (0.23)	0.54 (0.12)	8.78 (1.97)	0.50 (0.11)	0.60 (0.13)	10.1 (2.26)

Note

[a]Modified bitumen assemblies consist of one ply 3 mm smooth-surfaced base and one ply 4 mm granulated cap.

on 80 and 100 measurements, respectively. Using their respective standard deviations to achieve 95% confidence levels, total ply thickness increases to 7.34 mm. Again, there are some anomalies with these values, but in general, thickness of the interply asphalt layer decreased with increasing application temperature.

Chart B

	Roll-In				**Fly-In**				**Mop-and-Flop**		
	T.III	**T.IV**	**L**		**T. III**	**T.IV**	**L**		**T. III**	**T. IV**	**L**
218 °C	8.33	8.45	8.85		9.09	9.04	8.94		10.33	9.72	9.68
232 °C	8.27	8.22	8.22		8.64	8.68	8.49		9.58	9.92	9.68
246 °C	7.84	8.16	8.13		8.39	8.51	8.43		10.1	9.55	9.06
260 °C	7.95	7.82	8.06		8.60	8.67	8.63		9.38	9.35	9.65

Conclusions

Results from this work clearly indicate no gain in adhesion is realized with increasing application temperature. In fact, the elevated temperatures tend to thin interply mopping layers that will cool at an even faster rate. This becomes quite important in the field where ambient conditions of temperature and wind can be of significant effect. Obviously all of these conditions could exacerbate adhesion levels or the potential for blisters.

The authors agree that a more thoroughly bonded system should be a goal for improved service life and system durability. However, increasing application temperature is not the answer, particularly when considering other negative aspects attendant with elevated heating of bitumen. EVT and EVT range remain relevant concepts for application of asphalts in polymer modified membrane systems.

If improved adhesion is a primary criterion in polymer modified membrane systems, then the use of ASTM D 6152 SEBS modified mopping asphalts must be given consideration. The authors also understand that the use of ASTM D 312, Type IV asphalt has been used to address the potential for asphalt slippage. Based upon the decrease in adhesion with Type IV, this isn't necessarily a good tradeoff.

Currently within Committee D08 there exists a task group attempting to classify mopping asphalts through viscosity grading. Simply from the aspect of applying this concept to the polymer modified membrane system, this work could be invaluable and should be endorsed by a wider participation from the roofing community.

References

[1] Smith, T. L., "SBS-modified bitumen membranes: Tips for mop application," *Professional Roofing,* February 1996, p. 50.

[2] Baxter, D., "A Tale of Blisters," *Contractors Guide,* July 1998, pp. 24 – 31.

[3] National Roofing Contractors Association and National Research Council of Canada, *Cause of Blistering in SBS Polymer Modified-Bitumen Roofing Membranes,* March 1998.

[4] Graham, M. S., "Reports of Blistering are Increasing," *Professional Roofing,* April 2000, p. 55.

[5] Graham, M. S., "Minimizing SBS Blistering," *Professional Roofing,* May 2000, p. 71.

[6] Carlson, J., "Cap-sheet application recommendations," *Professional Roofing,* August 1999, pp. 29 – 32.

[7] Chernotowich, K., Brzozowski, K. J., and Fulcher, H. E., "Effect of Application Temperature on Adhesion of Modified Bitumen Membrane Assemblies," *8th International Conference on Durability of Building Materials and Components (8dbmc), Paper No. 102,* Canada 1999.

Karen Liu,[1] *Guodong Xu,*[2] *and Richard Voyer*[3]

Durability and Cold Temperature Performance of SBS-Modified Bituminous Roofing Membranes

REFERENCE: Liu, K., Xu, G., and Voyer, R., "**Durability and Cold Temperature Performance of SBS-Modified Bituminous Roofing Membranes,**" *Roofing Research and Standards Development: 5th Volume, ASTM STP 1451*, W. J. Rossiter and T. J. Wallace., Eds., ASTM International, West Conshohocken, PA, 2003.

ABSTRACT: The National Research Council of Canada, in collaboration with a Canadian roofing manufacturer, conducted an experimental program to evaluate the durability of SBS-modified bituminous roofing membranes in Canada. This study examined the effect of heat aging on the mechanical and chemical properties of nine commercially available SBS-modified membranes, with emphasis on cold temperature performance. These membranes varied in reinforcements and densities, and were selected to be representative of SBS-modified membranes currently used in Canada. The membrane samples were heat aged in convection ovens at 70±3°C for six months. During this process, some samples were withdrawn at two-month intervals and their tensile properties at 23°C and -30°C, cold flex temperature and glass transition temperature determined. The results confirmed that heat aging mainly affected matrix-controlled properties of the membranes. In addition, the reinforcement controlled the mechanical properties at 23°C but the modified bitumen governed the failure mechanism at –30°C for the polyester reinforced membranes.

KEYWORDS: SBS, modified bituminous membrane, durability, tensile, strength, elongation, strain energy, low temperature flexibility, cold flex temperature, glass transition temperature, failure mechanism, low temperature, cold weather

Introduction

There is nowhere in Canada where the January design temperature is above freezing [*1*]. Because of our cold weather, it is important that roofing membranes perform well at low temperature. Roofing membranes should remain flexible at low temperature to resist stresses experienced on the roofs. Most single-ply roofing membranes are formulated with plasticizers to enhance low temperature flexibility. Polymer modifiers such as styrene-butadiene-styrene (SBS) are added to bitumen to improve elongation and flexibility. SBS-modified bituminous membranes are commonly used in Canada because of their low temperature performance.

Accelerated weathering is often performed on roof membranes to accelerate the effects of natural weathering. Although changes in physical and chemical properties due to accelerated weathering might not directly correlate to those occurring naturally, nevertheless, accelerated weathering using techniques such as heat aging, QUV and

[1] Research Officer, National Research Council, 1200 Montreal Road, Ottawa, ON, K1A 0R6.

[2] Industrial Research Fellow, National Research Council, 1200 Montreal Road, Ottawa, ON, K1A 0R6.

[3] Research Director, Soprema Inc., 1675 Rue Haggerty, Drummondville, Quebec, J2C 5P7.

Xenon-arc provides useful information in evaluating the durability of roof membranes. Heat aging is probably the most commonly used accelerated weathering technique in studying durability of roofing membranes as it can be easily performed in the laboratory. Heat aging has been reported to be effective in evaluating the performance of modified bituminous membrane [*2*] and simulating the effects of natural aging [*3*]. Aging of bitumen is believed to be a combination of the migration of the light volatile components and the chemical changes produced by the effects of oxygen, energy and water [*4*]. The heat aging process intensifies the chemical changes by increasing the rates of chemical reactions involved in the aging process and thus simulates at a faster rate the aging mechanisms which occur naturally.

Puterman et al. [*5*] measured the tensile properties, the cold flex temperature and the water-pressure resistance of various types of roofing membrane which were naturally exposed under normal service conditions. Data on the SBS- and APP-modified membranes showed that the exposure hardly affected the properties determined by the reinforcement but had a strong effect on the properties that are governed by the bitumen/polymer material. In particular, they reported that the cold flex temperature of a SBS-modified bituminous membrane increased by about 12°C after five years of natural exposure but the tensile strength and elongation were hardly affected. The data also showed that SBS-modified membranes retained better low-temperature flexibility than the APP-modified membranes, after comparable periods of exposure under similar service conditions.

Baxter et al. [*6*] showed that the tensile strength of five SBS-modified bituminous membranes increased and the cold flex temperature of two SBS-modified bituminous membranes rose by about 15°C, after heat aging at 80°C for 28 days. Rodriguez et al. [*7*] studied the effects of heat aging and test temperature on the tensile strength and elongation of two APP-modified and two SBS-modified bituminous membranes. For the two SBS-modified membranes, heat aging (80°C for 168 days) did not significantly affect the tensile strength at 23°C but it reduced the elongation by 20-40%. However, heat aging reduced the tensile strength of these membranes at –30°C by about 20% and increased the elongation by about 60-330%. They also found that the glass transition temperature of these membranes increased by 4 – 8°C.

This study systematically examined the durability of nine SBS-modified bituminous roofing membranes by heat aging. The primary objective was to quantify the effects of heat aging on the physical and mechanical properties of these membranes, with particular emphasis on the cold temperature performance. The membranes were selected to be representative of SBS-modified currently used in Canada; therefore, the data reflects the current status of SBS-modified membranes in Canada. A secondary objective was to gain a better fundamental understanding of the failure mechanism of SBS-modified membranes at low temperature, which is important in improving membrane performance for cold weather applications.

Experimental Method

Materials and Heat Aging

Nine commercially available SBS-modified bituminous roofing membrane samples were used in this study. Three rolls were obtained from each of three manufacturers (A, B and C) through normal distribution channels. From the three rolls, one roll was reinforced with random short glass fiber mat and the other two with a non-woven polyester mat of two different nominal reinforcement densities (180 g·m^{-2} and 250 g·m^{-2}). The top surface of the membranes reinforced with the 180 g·m^{-2} mat was smooth but that of the membranes reinforced with the 250 g·m^{-2} mat was covered with granules. The glass fiber reinforced membrane was covered with a thin layer of sand. Table 1 provides a summary of the samples used in this study.

Samples were randomly selected, cut from the rolls and placed in convection ovens at 70±3°C for six months. During this process, some of the samples were withdrawn at two-month intervals (i.e. 0, 2, 4 and 6 months) for physical and mechanical tests.

Table 1 – *Description of the SBS-modified bituminous membranes used in this study.*

Sample ID	Manu-facturer	Reinforcement Type	Nominal Reinforcement Density (g·m^{-2})	Membrane Thickness (mm)	Surface Characteristics
A1	A	Glass	Not Available	2.1	Sand
B1	B	Glass	Not Available	2.2	Sand
C1	C	Glass	Not Available	2.0	Sand
A2	A	Polyester	180	3.1	Smooth
B2	B	Polyester	180	3.0	Smooth
C2	C	Polyester	180	3.1	Smooth
A3	A	Polyester	250	4.2	Granules
B3	B	Polyester	250	4.2	Granules
C3	C	Polyester	250	4.3	Granules

Testing

Tensile Properties - The tensile properties were measured according to CGSB-37.56M Modified Bituminous Membranes for Roofing [*8*]. The specimens (25 mm X 150 mm) were cut from the membrane samples, in both machine (MD) and cross-machine (XD) directions, using a rectangular die in a hydraulic press. They were tested using an Instron 4502 Automated Materials Testing System connected to a computer for data recording. The grip distance was 75 mm. A minimum of five specimens was tested at each condition. The breaking strength, the elongation at break and the strain energy to ultimate elongation were measured (definitions of these parameters are described in the CGSB-37.56M standard).

For the room temperature tests, the specimens were placed in the hydraulic grips and pulled at a constant crosshead speed of 50 mm·min^{-1}. A 1-kN load cell was used to monitor the applied load. For the tests performed at –30°C, the specimens were pre-cooled at –30°C in a laboratory freezer overnight. An Instron environmental chamber with liquid nitrogen cooling was used to maintain the cold temperature of the specimens during testing. The specimens were pulled at a constant crosshead speed of 10 mm·min^{-1}. A 10-kN load cell was used to monitor the applied load. Because hydraulic grips could not be used at the low temperatures, the specimen ends were sandwiched between a pair of specially designed tabs that could be tightened with two screws. This specimen-tab assembly was slid into an accompanying fixture (Figure 1) which was mounted in the testing machine. The specimens were allowed to equilibrate in the environmental chamber at –30°C for at least 5 minutes before testing.

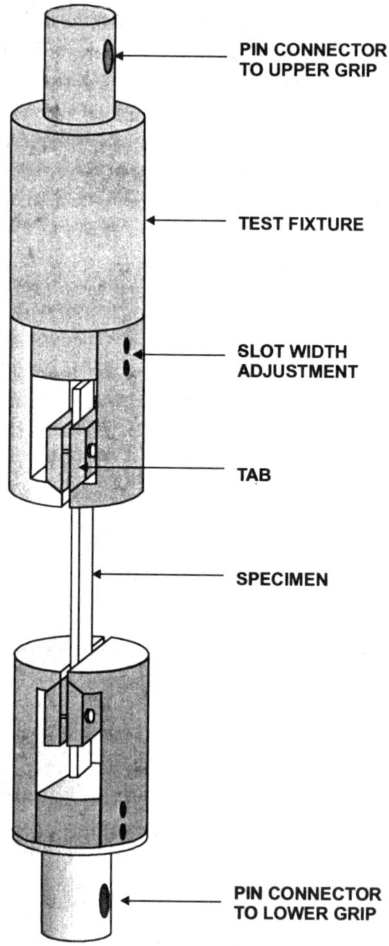

Figure 1 – *Loading assembly for tensile tests at cold temperatures.*

Low Temperature Flexibility - The low temperature flexibility was determined according to ASTM D 5147-97 Standard Test Methods for Sampling and Testing Modified Bituminous Sheet Material. The specimens (25 mm X 150 mm) were cut from the membrane samples, in both machine (MD) and cross-machine (XD) directions, using a rectangular die in a hydraulic press. A minimum of five specimens was cut from each of the control and heat-aged membranes in both MD and XD directions. The specimens and the mandrel (25 mm in diameter) were allowed to equilibrate in an upright freezer, at the test temperature, for three hours before testing. The specimen was bent 180±5° around the mandrel at the test temperature in 2 ± 1 s. If a crack was observed in any of the five specimens, the test temperature was increased by 3°C; however, if no crack was observed in any of the specimens, the test temperature was decreased by 3°C. The specimens and the mandrel were allowed to equilibrate for another 30 minutes after the freezer reached the new test temperature before the next round of testing. The test temperature was adjusted until the lowest temperature at which none of the specimens showed any visual signs of cracking was determined, and this temperature was recorded as the cold flex temperature (T_{CF}).

Glass Transition Temperature - The glass transition temperature of the samples was measured according to the ASTM D6382-99 Standard Practice for Dynamic Mechanical Analysis and Thermogravimetry of Roofing and Waterproofing Membrane Material and D5418-99 Standard Test Method for Plastics: Dynamic Mechanical Properties: in Flexure (Dual Cantilever Beam) using a dynamic mechanical thermal analyzer Rheometrics Scientific Solids Analyzer RSA II. The control and heat-aged membranes were cut in the machine direction (3 mm X 37 mm) and tested in the dual cantilever mode. The granules on the top surface of the membranes were removed as much as possible. A fixed oscillation frequency of 1 Hz and a heating rate of 2°C·min^{-1}in the range of –80°C to 30°C were used. The instrument measured the storage modulus (E') and the loss modulus (E") of the specimen. The glass transition temperature was obtained for the maximum in the loss modulus (E") versus temperature curve. At least two specimens were tested for each sample.

Results and Discussion

Tensile Properties

Force-Displacement Curves - Typical force-displacement curves recorded by the Instron testing machine are shown in Figures 2a, 3a and 4a. For the glass fiber reinforced membranes (A1, B1 and C1), the specimens generally failed abruptly in the elastic region with a single crack (Figure 2a). For Sample A1 tested at 23°C, the bituminous matrix was flexible and stretched a great deal after the reinforcement was broken. However, the bituminous matrix lost this flexibility at –30 °C and the specimen failed with a clean single crack. There was a distinct "knee" on the force-displacement curve of Sample C1 tested at –30°C where the modulus (slope of the curve) changed. This could be due to the change in modulus at matrix yield (more significant at low temperature) or change in reinforcement structure (e.g. fiber sliding).

The polyester reinforced membranes had much higher elongation at break than those reinforced with glass fiber due to the high elongation characteristics of the polymeric reinforcement (Figures 3a and 4a). The membrane behaved in an elastic-plastic manner with a change in modulus at the yield point, at both 23°C and –30°C. Heat aging did not affect the maximum tensile strength but it decreased the elongation considerably at 23°C. All membranes reinforced at 250 g·m^{-2} (A3, B3 and C3) failed with a single crack but some of those reinforced at 180 g·m^{-2} (A2, B2 and C2) did not. The control of Sample A2 failed with a single crack at -30°C after being extended to 25 mm but the heat-aged specimens failed after extending to only about 5 mm. Several small cracks were

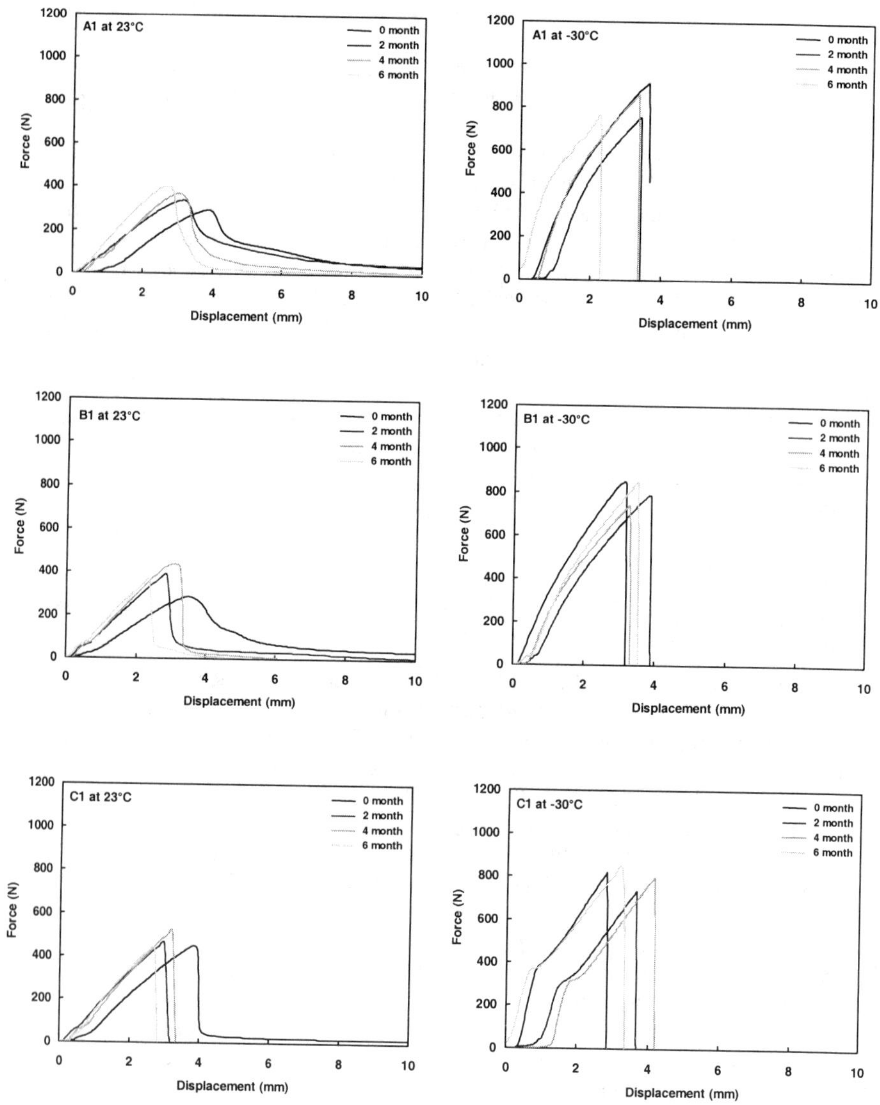

Figure 2a – Typical tensile force-displacement curves of SBS-modified membranes (reinforced with random short glass fiber mat).

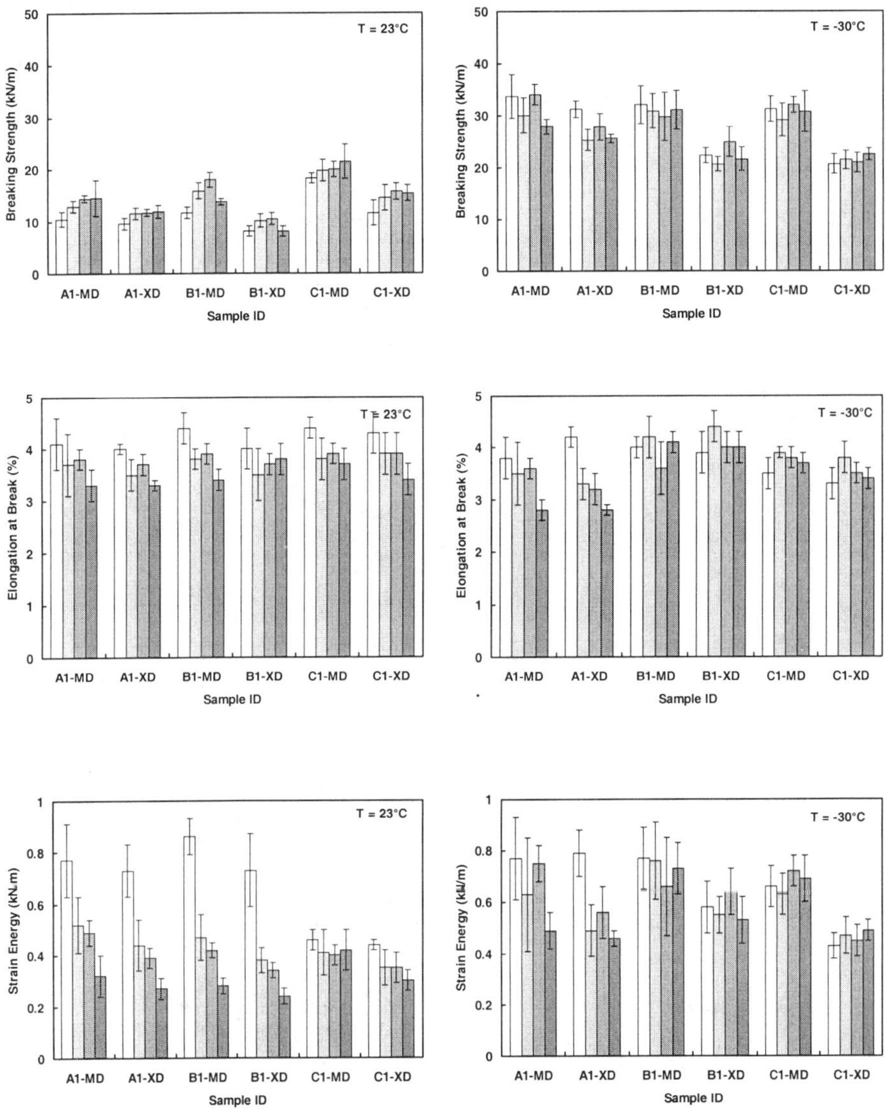

Figure 2b – Effects of heat aging on the tensile properties of SBS-modified membranes (reinforced with random short glass fiber mat).

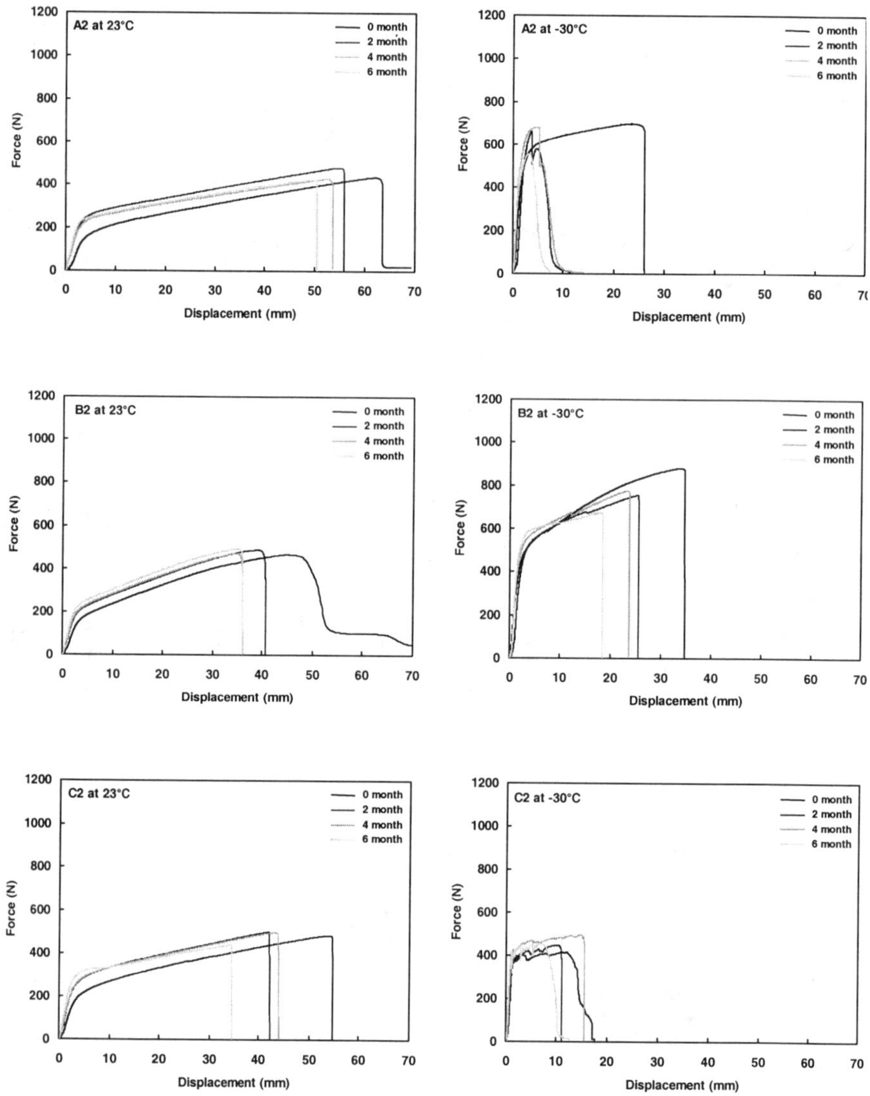

Figure 3a – *Typical tensile force-displacement curves of SBS-modified membranes (reinforced with polyester mat at 180 g·m^{-2}).*

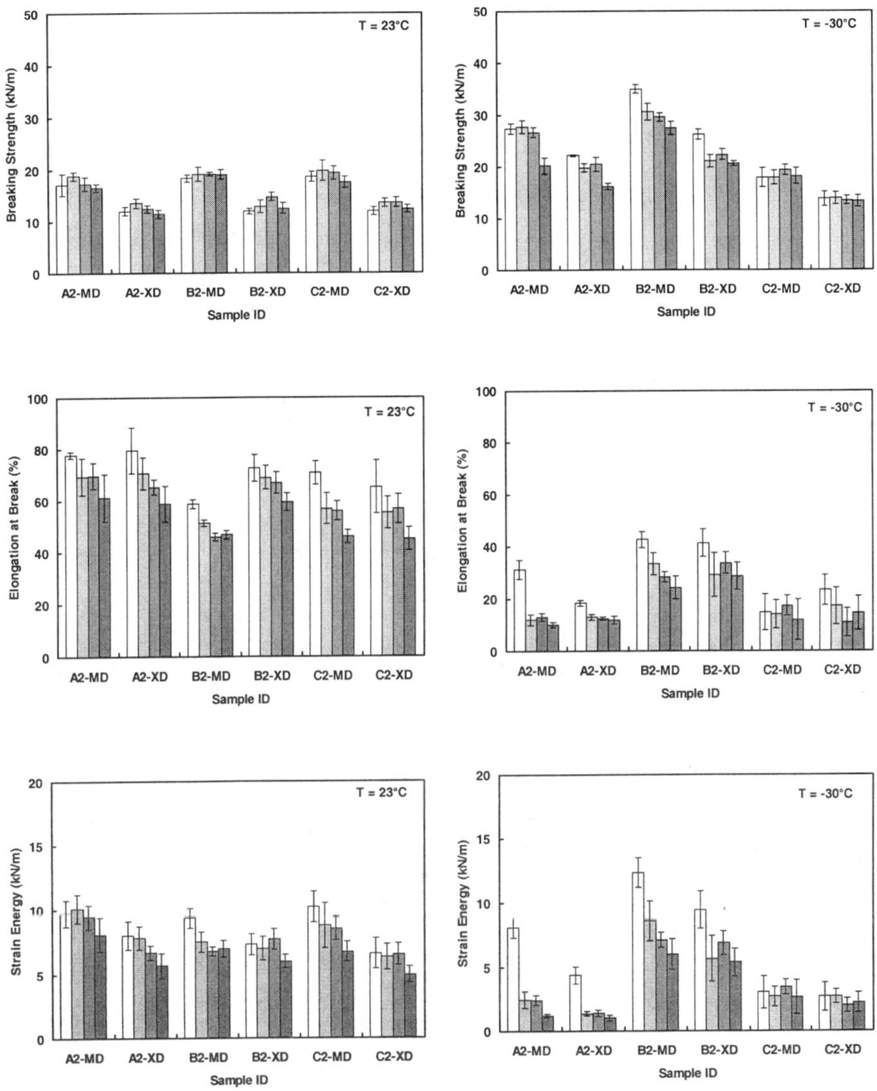

Figure 3b – *Effects of heat aging on the tensile properties of SBS-modified membranes (reinforced with polyester mat at 180 g·m^{-2}).*

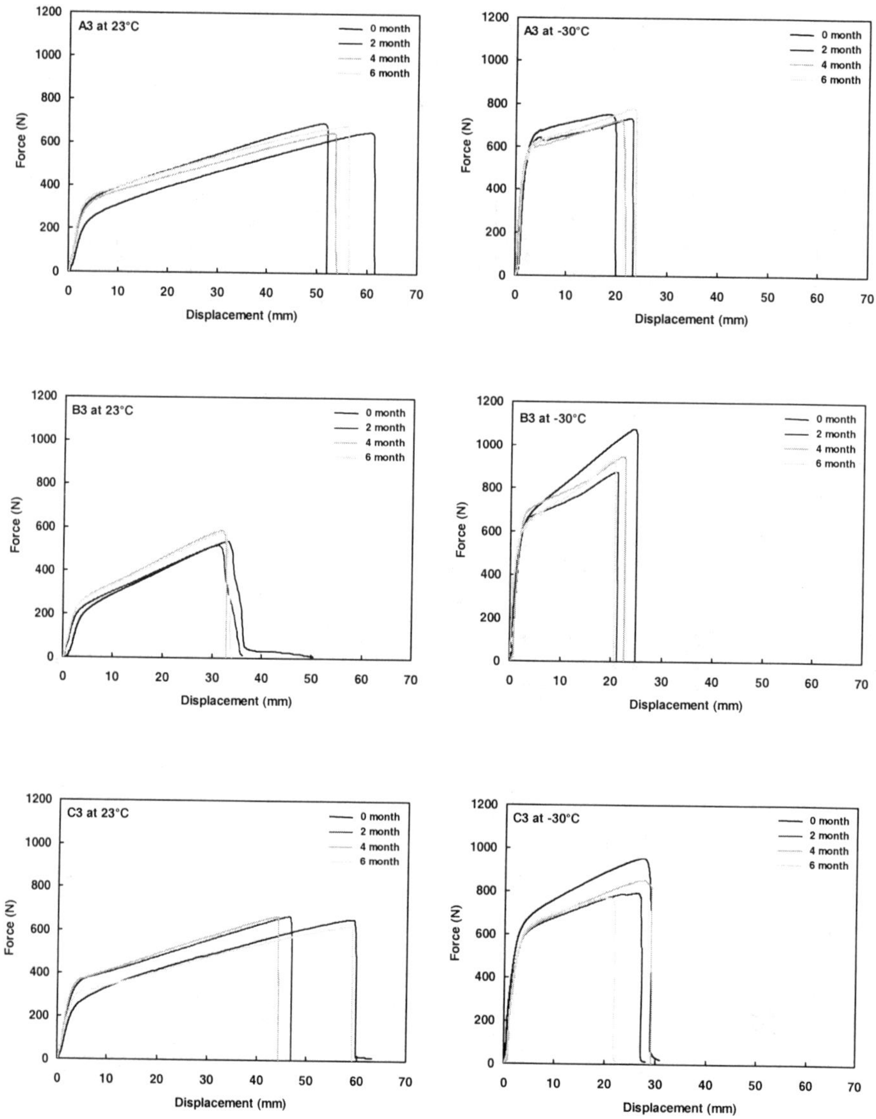

Figure 4a – *Typical tensile force-displacement curves of SBS-modified membranes (reinforced with polyester mat at 250 g·m^{-2}).*

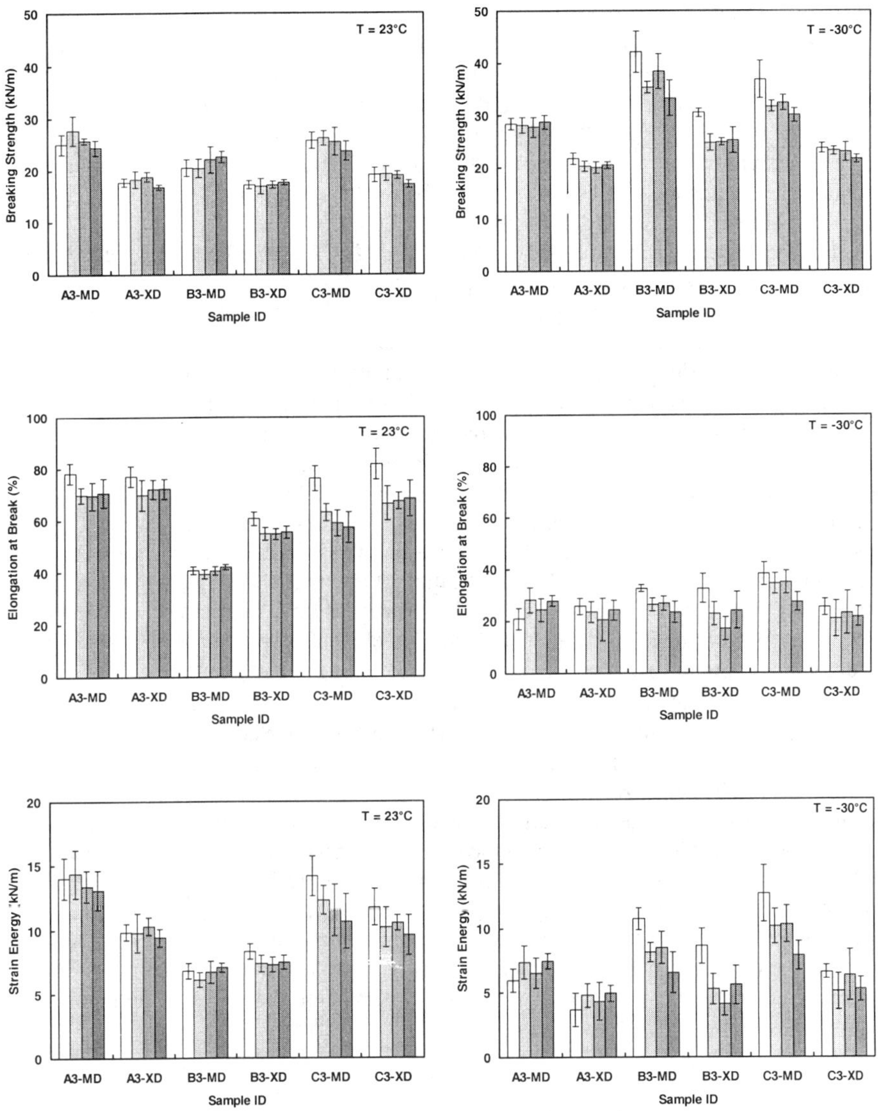

Figure 4b – *Effects of heat aging on the tensile properties of SBS-modified membranes (reinforced with polyester mat at 250 g·m⁻²).*

observed to form on the heat-aged bituminous matrix of Sample A2 at –30°C before it fractured. For Sample C2, multiple small cracks were observed to form and grow on the bituminous matrix, and final fracture occurred at one of the larger cracks.

Effect of Heat Aging - The tensile properties of the membranes at different test temperatures and aging periods are summarized in Tables 2-4 and Figures 2b, 3b and 4b. The values reported are the average and standard deviation of at least five specimens. For the glass fiber reinforced membranes (A1, B1 and C1), heat aging did not affect the breaking strength but reduced the elongation at break and the strain energy slightly (Figure 2b). Samples A1 was affected most by the heat aging: its strain energy dropped by about 60% at 23°C and 40% at –30°C. Sample C1 was least affected.

For the polyester reinforced membranes (A2, B2, C2, A3, B3 and C3), heat aging did not affect the breaking strength of the membranes at 23°C but it reduced the elongation at break by as much as 35% (Figures 3b and 4b). The reduction in elongation was more severe in the membranes containing less reinforcement: 20 to 35% for the 180 $g{\cdot}m^{-2}$ compared to 0 to 25% for the 250 $g{\cdot}m^{-2}$ over six months of heat aging. Because of the reduced elongation, the strain energy was also decreased accordingly (15-35% for the 180 $g{\cdot}m^{-2}$ and 0-25% for the 250 $g{\cdot}m^{-2}$ over six months). Note that the elongation at break and strain energy of Sample A2 at –30°C was reduced significantly after heat aging due to the change in failure mode as shown by the force-displacement curves.

The effects of heat aging (reduction in elongation and strain energy) were generally more prominent at –30°C. This is because the matrix shared a larger portion of the load at low temperature and governed the elongation of the membrane at low temperature, and heat aging mainly affected matrix-controlled properties (this will be discussed in the next section). As with the 23°C data, the reduction in elongation was more pronounced in the membranes containing less reinforcement (20-70% for the 180 $g{\cdot}m^{-2}$ and 0-30% for the 250 $g{\cdot}m^{-2}$ over six months). It is interesting to note that, within experimental errors, the tensile properties (strength, elongation and strain energy) of Sample A3 were hardly affected by six months of heat aging, indicating its stability.

Effects of Temperature on Failure Mechanism - A roofing membrane is a composite material with a multi-layer structure in which a reinforcement is embedded in a matrix. The reinforcement bears the load and the matrix binds the reinforcement together and allows stress transfer between the two components. For example, with the polyester reinforced membranes in this study, the non-woven polyester mat is the reinforcement and the modified bitumen is the matrix. The failure of the membrane can be described by the mechanics of composite materials. Rule of Mixture (ROM) describes the strength of a composite material with aligned, continuous fibers [9] in the direction along the fibers:

$$\sigma_c = (V_f{\cdot}E_f + V_m{\cdot}E_m)\,\varepsilon_c \tag{1}$$

where

σ_c	=	composite stress ($N{\cdot}m^{-2}$)
E_f	=	fiber modulus ($N{\cdot}m^{-2}$)
E_m	=	matrix modulus ($N{\cdot}m^{-2}$)
V_f	=	fiber volume fraction (dimensionless)
V_m	=	matrix volume fraction (dimensionless)
ε_c	=	composite strain (dimensionless)

Table 2 – *Effects of heat aging on the tensile breaking strength.*

Test Temp.	Sample ID	Test Dir.	Breaking Strength* ($kN \cdot m^{-1}$) after heat aging at 70°C for:			
			0 month	2 months	4 months	6 months
23°C	A1	MD	10.5 ± 1.4	12.9 ± 1.1	14.4 ± 0.7	14.5 ± 3.4
		XD	9.6 ± 1.1	11.5 ± 1.1	11.7 ± 0.7	11.9 ± 1.2
	B1	MD	11.7 ± 1.1	15.9 ± 1.5	18.0 ± 1.4	13.8 ± 0.6
		XD	8.1 ± 1.0	10.1 ± 1.3	10.5 ± 1.2	8.1 ± 1.0
	C1	MD	18.3 ± 1.0	19.8 ± 2.1	20.0 ± 1.5	21.5 ± 3.3
		XD	11.6 ± 2.4	14.5 ± 2.5	15.7 ± 1.6	15.4 ± 1.5
	A2	MD	17.1 ± 2.1	18.8 ± 0.8	17.3 ± 1.3	16.5 ± 0.7
		XD	12.0 ± 0.8	13.5 ± 0.9	12.4 ± 0.7	11.4 ± 0.8
	B2	MD	18.3 ± 0.7	19.1 ± 1.4	19.1 ± 0.4	19.0 ± 0.9
		XD	11.9 ± 0.6	12.8 ± 1.2	14.7 ± 0.8	12.5 ± 1.0
	C2	MD	18.6 ± 1.0	19.7 ± 2.0	19.3 ± 1.3	17.5 ± 1.1
		XD	11.9 ± 0.8	13.5 ± 0.8	13.6 ± 0.9	12.3 ± 0.7
	A3	MD	25.0 ± 1.9	27.7 ± 2.8	25.7 ± 0.6	24.4 ± 1.5
		XD	17.7 ± 0.8	18.3 ± 1.6	18.8 ± 0.9	16.8 ± 0.5
	B3	MD	20.5 ± 1.6	20.4 ± 1.8	22.0 ± 2.5	22.6 ± 1.1
		XD	17.2 ± 0.8	16.9 ± 1.5	17.2 ± 0.7	17.6 ± 0.5
	C3	MD	25.7 ± 1.6	26.2 ± 1.4	25.5 ± 2.7	23.7 ± 1.9
		XD	19.1 ± 1.4	19.3 ± 1.4	19.0 ± 0.7	17.3 ± 0.7
-30°C	A1	MD	33.7 ± 4.2	30.1 ± 3.4	34.1 ± 2.0	27.9 ± 1.4
		XD	31.3 ± 1.6	25.3 ± 2.0	27.8 ± 2.6	25.6 ± 0.8
	B1	MD	32.1 ± 3.6	30.9 ± 3.3	29.8 ± 4.7	31.1 ± 3.7
		XD	22.3 ± 1.5	20.6 ± 1.4	24.9 ± 2.9	21.6 ± 2.3
	C1	MD	31.3 ± 2.5	29.2 ± 3.2	32.1 ± 1.5	30.8 ± 3.9
		XD	20.6 ± 1.9	21.4 ± 1.8	20.9 ± 1.9	22.5 ± 1.2
	A2	MD	27.4 ± 1.0	27.8 ± 1.2	26.7 ± 0.9	20.2 ± 1.5
		XD	22.2 ± 0.2	19.8 ± 0.8	20.5 ± 1.4	16.2 ± 0.6
	B2	MD	35.0 ± 0.8	30.6 ± 1.6	29.6 ± 0.8	27.5 ± 1.3
		XD	26.3 ± 1.0	21.1 ± 1.2	22.4 ± 1.0	20.6 ± 0.5
	C2	MD	17.9 ± 1.8	17.9 ± 1.3	19.4 ± 1.0	18.2 ± 1.5
		XD	13.8 ± 1.4	13.9 ± 1.2	13.5 ± 0.8	13.3 ± 1.1
	A3	MD	28.5 ± 1.1	28.2 ± 1.5	27.8 ± 1.9	28.8 ± 1.3
		XD	21.7 ± 1.1	20.2 ± 1.0	19.9 ± 1.1	20.4 ± 0.7
	B3	MD	42.1 ± 3.9	35.3 ± 1.1	38.4 ± 3.3	33.3 ± 3.4
		XD	30.5 ± 0.8	24.8 ± 1.6	24.9 ± 0.7	25.2 ± 2.5
	C3	MD	36.8 ± 3.6	31.7 ± 1.0	32.4 ± 1.5	30.0 ± 1.3
		XD	23.6 ± 1.0	23.1 ± 0.8	22.9 ± 1.9	21.5 ± 0.8

* average ± standard deviation of at least 5 specimens

Table 3 – *Effects of heat aging on the tensile elongation at break.*

Test Temp.	Sample ID	Test Dir.	Elongation at Break* (%) after heat aging at 70°C for: 0 month	2 months	4 months	6 months
23°C	A1	MD	4.1 ± 0.5	3.7 ± 0.6	3.8 ± 0.2	3.3 ± 0.3
		XD	4.0 ± 0.1	3.5 ± 0.3	3.7 ± 0.2	3.3 ± 0.1
	B1	MD	4.4 ± 0.3	3.8 ± 0.2	3.9 ± 0.2	3.4 ± 0.2
		XD	4.0 ± 0.4	3.5 ± 0.5	3.7 ± 0.2	3.8 ± 0.3
	C1	MD	4.4 ± 0.2	3.8 ± 0.4	3.9 ± 0.2	3.7 ± 0.3
		XD	4.3 ± 0.4	3.9 ± 0.4	3.9 ± 0.4	3.4 ± 0.3
	A2	MD	77.7 ± 1.2	69.4 ± 7.2	69.7 ± 5.1	61.2 ± 9.1
		XD	79.6 ± 8.9	70.6 ± 6.2	65.3 ± 2.9	58.7 ± 6.9
	B2	MD	58.7 ± 1.7	51.4 ± 1.2	45.9 ± 1.4	46.9 ± 1.6
		XD	72.6 ± 5.1	69.0 ± 4.5	66.9 ± 4.3	59.5 ± 3.4
	C2	MD	70.8 ± 4.6	56.9 ± 6.1	56.1 ± 3.7	46.1 ± 2.6
		XD	65.3 ± 10.3	55.4 ± 6.2	56.8 ± 5.6	45.2 ± 4.4
	A3	MD	78.2 ± 3.9	69.9 ± 3.0	69.6 ± 5.3	70.7 ± 5.5
		XD	77.1 ± 3.9	69.9 ± 5.9	72.1 ± 3.8	72.3 ± 3.8
	B3	MD	40.8 ± 1.5	39.3 ± 1.8	40.6 ± 1.6	41.9 ± 1.0
		XD	60.7 ± 2.6	54.8 ± 2.4	54.7 ± 2.0	55.5 ± 2.3
	C3	MD	76.2 ± 4.8	63.1 ± 3.4	58.9 ± 5.0	57.3 ± 5.9
		XD	81.8 ± 5.9	66.5 ± 6.6	67.4 ± 3.1	68.3 ± 6.9
-30°C	A1	MD	3.8 ± 0.4	3.5 ± 0.6	3.6 ± 0.2	2.8 ± 0.2
		XD	4.2 ± 0.2	3.3 ± 0.3	3.2 ± 0.3	2.8 ± 0.1
	B1	MD	4.0 ± 0.2	4.2 ± 0.4	3.6 ± 0.5	4.1 ± 0.2
		XD	3.9 ± 0.4	4.4 ± 0.3	4.0 ± 0.3	4.0 ± 0.3
	C1	MD	3.5 ± 0.3	3.9 ± 0.1	3.8 ± 0.2	3.7 ± 0.2
		XD	3.3 ± 0.3	3.8 ± 0.3	3.5 ± 0.2	3.4 ± 0.2
	A2	MD	31.1 ± 3.7	11.9 ± 2.1	13.0 ± 1.5	10.1 ± 0.9
		XD	18.4 ± 1.1	12.9 ± 1.2	12.4 ± 0.6	11.8 ± 1.4
	B2	MD	42.7 ± 3.1	33.3 ± 4.4	28.2 ± 1.9	24.2 ± 4.3
		XD	41.4 ± 5.4	29.1 ± 8.5	33.7 ± 4.2	28.6 ± 5.3
	C2	MD	14.8 ± 6.9	14.0 ± 5.4	17.3 ± 3.8	11.9 ± 7.8
		XD	23.3 ± 5.8	17.1 ± 7.0	10.7 ± 5.4	14.4 ± 6.5
	A3	MD	21.0 ± 4.1	28.2 ± 4.9	24.5 ± 4.4	27.9 ± 2.2
		XD	25.7 ± 3.0	23.6 ± 4.0	20.6 ± 8.1	24.2 ± 3.9
	B3	MD	32.6 ± 1.3	26.2 ± 2.5	26.8 ± 2.8	23.4 ± 4.1
		XD	32.6 ± 5.6	22.8 ± 4.4	17.1 ± 4.4	24.1 ± 7.1
	C3	MD	38.2 ± 4.5	34.4 ± 4.0	35.0 ± 4.5	27.4 ± 3.5
		XD	25.3 ± 3.3	20.8 ± 7.0	23.2 ± 8.3	21.7 ± 3.8

* average ± standard deviation of at least 5 specimens

Table 4 – *Effects of heat aging on the tensile strain energy.*

Test Temp.	Sample ID	Test Dir.	Strain Energy* (kN·m^{-1}) after heat aging at 70°C for:			
			0 month	2 months	4 months	6 months
23°C	A1	MD	0.77 ± 0.14	0.52 ± 0.11	0.49 ± 0.05	0.32 ± 0.08
		XD	0.73 ± 0.10	0.44 ± 0.10	0.39 ± 0.04	0.27 ± 0.04
	B1	MD	0.86 ± 0.07	0.47 ± 0.09	0.42 ± 0.03	0.28 ± 0.03
		XD	0.73 ± 0.14	0.38 ± 0.05	0.34 ± 0.03	0.24 ± 0.03
	C1	MD	0.46 ± 0.04	0.41 ± 0.09	0.40 ± 0.04	0.42 ± 0.08
		XD	0.44 ± 0.02	0.35 ± 0.07	0.35 ± 0.06	0.30 ± 0.04
	A2	MD	9.73 ± 1.02	10.10 ± 1.08	9.43 ± 0.91	8.08 ± 1.30
		XD	8.01 ± 1.10	7.83 ± 0.89	6.68 ± 0.54	5.67 ± 0.97
	B2	MD	9.33 ± 0.76	7.50 ± 0.78	6.76 ± 0.35	6.98 ± 0.65
		XD	7.29 ± 0.81	6.98 ± 0.93	7.72 ± 0.80	5.98 ± 0.52
	C2	MD	10.21 ± 1.19	8.76 ± 1.73	8.53 ± 0.93	6.74 ± 0.76
		XD	6.58 ± 1.17	6.30 ± 0.99	6.52 ± 0.84	4.95 ± 0.62
	A3	MD	14.00 ± 1.59	14.34 ± 1.87	13.39 ± 1.23	13.07 ± 1.52
		XD	9.85 ± 0.63	9.77 ± 1.49	10.29 ± 0.68	9.41 ± 0.67
	B3	MD	6.81 ± 0.61	6.10 ± 0.58	6.72 ± 0.86	7.09 ± 0.35
		XD	8.33 ± 0.62	7.39 ± 0.64	7.30 ± 0.56	7.48 ± 0.55
	C3	MD	14.17 ± 1.54	12.28 ± 1.13	11.53 ± 1.99	10.64 ± 2.11
		XD	11.73 ± 1.42	10.16 ± 1.54	10.54 ± 0.60	9.58 ± 1.54
-30°C	A1	MD	0.77 ± 0.16	0.63 ± 0.22	0.75 ± 0.07	0.49 ± 0.07
		XD	0.79 ± 0.09	0.49 ± 0.10	0.56 ± 0.10	0.46 ± 0.03
	B1	MD	0.77 ± 0.12	0.76 ± 0.15	0.66 ± 0.19	0.73 ± 0.10
		XD	0.58 ± 0.10	0.55 ± 0.07	0.64 ± 0.09	0.53 ± 0.09
	C1	MD	0.66 ± 0.08	0.63 ± 0.08	0.72 ± 0.06	0.69 ± 0.09
		XD	0.43 ± 0.05	0.47 ± 0.07	0.45 ± 0.06	0.49 ± 0.04
	A2	MD	8.07 ± 0.79	2.46 ± 0.64	2.41 ± 0.37	1.22 ± 0.11
		XD	4.37 ± 0.69	1.37 ± 0.16	1.39 ± 0.24	1.03 ± 0.24
	B2	MD	12.30 ± 1.15	8.57 ± 1.54	7.10 ± 0.57	6.00 ± 1.18
		XD	9.41 ± 1.44	5.66 ± 1.78	6.87 ± 0.92	5.37 ± 1.07
	C2	MD	3.02 ± 1.26	2.67 ± 0.77	3.43 ± 0.61	2.66 ± 1.35
		XD	2.67 ± 1.14	2.67 ± 0.55	1.99 ± 0.53	2.20 ± 0.77
	A3	MD	5.97 ± 0.92	7.40 ± 1.28	6.56 ± 1.18	7.50 ± 0.61
		XD	3.67 ± 1.30	4.79 ± 0.92	4.34 ± 1.48	4.94 ± 0.64
	B3	MD	10.71 ± 0.83	8.13 ± 0.76	8.51 ± 1.24	6.55 ± 1.61
		XD	8.63 ± 1.36	5.26 ± 1.20	4.14 ± 0.91	5.60 ± 1.51
	C3	MD	12.64 ± 2.18	10.12 ± 1.32	10.28 ± 1.41	7.87 ± 1.13
		XD	6.62 ± 0.54	5.09 ± 1.42	6.36 ± 1.97	5.24 ± 0.94

* average ± standard deviation of at least 5 specimens

In commercial production, the fibers might not be aligned or continuous (e.g randomly oriented short fiber mat) and the strength of the composite usually reaches only a fraction of the rule of mixture strength [*10*]. The Rule of Mixture can be modified by taking the fiber orientation and length into account:

$$\sigma_c = (\chi_1 \chi_2 V_f \cdot E_f + V_m \cdot E_m)\, \varepsilon_c \quad (1a)$$

where

χ_1 = fiber orientation factor (dimensionless)
χ_2 = fiber length factor (dimensionless)

Both factors are less than or equal to unity. χ_1 is less than one when the fibers are aligned at an angle to the loading axis, and becomes unity when all fibers are aligned to the loading axis. Similarly, χ_2 is less than one when the fibers are short, and becomes unity when the fiber length is long and continuous (above the critical aspect ratio [*9*]).

At 23°C, the modulus of the polyester reinforcement was much higher than that of the modified bitumen (i.e. $E_f \gg E_m$). Therefore, the polyester fibers bore most of the load exerted on the membrane. The membrane failed when the polyester mat broke (i.e. $\varepsilon_c = \varepsilon_{fu}$). In this situation, equation (1a) becomes:

$$\sigma_{cu} = \chi_1 \chi_2 V_f \cdot \sigma_{fu} \quad (2)$$

where

σ_{cu} = failure strength of composite ($N \cdot m^{-2}$)
ε_{fu} = failure strain of fiber (dimensionless)
σ_{fu} = failure strength of fiber ($N \cdot m^{-2}$)

Since the polyester mat governed the tensile properties of the membrane at 23°C, as described by equation (2), improvement of mechanical performance of the membrane at room temperature can be achieved by using good quality polyester fiber (high strength and elongation properties) and/or increasing the amount of reinforcement.

In the roofing industry, the strength of the membrane containing reinforcement is generally expressed in force per unit length ($N \cdot m^{-1}$) because the membrane is always installed in plys. We can rewrite equation (2):

$$S_{cu} = \chi_1 \chi_2 a_f \cdot \sigma_{fu} / \rho_f \quad (2a)$$

where

S_{cu} = membrane strength per unit length ($N \cdot m^{-1}$)
a_f = aeral reinforcement density ($g \cdot m^{-2}$)
ρ_f = volumetric reinforcement density ($g \cdot m^{-3}$)

Equation (2a) indicates that S_{cu} is proportional to a_f, i.e., the strength of the membrane increases with the amount of reinforcement given that the same type of reinforcement is used. The strength ratio of the membranes reinforced at 180 $g \cdot m^{-2}$ and 250 $g \cdot m^{-2}$ of polyester mat were computed (i.e. strength of Sample A2 / strength of Sample A3) for each manufacturer (A, B and C) at both loading directions (MD and XD). The overall ratio was determined to be 0.74±0.08. This experimental value corresponded well with the theoretical strength ratio of 0.72 (180/250).

As the test temperature dropped, the modulus of the modified bitumen increased and the matrix began to carry an increasing proportion of the load exerted on the membrane ($V_m \cdot E_m \cdot \varepsilon_c$). Since the reinforcement and the matrix bore the load together, the strength of the membrane at –30°C was higher than that at 23°C. However, the ultimate elongation of the matrix reduced greatly as the temperature dropped. When the matrix started to crack (i.e. $\varepsilon_c = \varepsilon_{mu} < \varepsilon_{fu}$), equation (1) becomes:

$$\sigma_c = (\chi_1 \chi_2 V_f E_f + V_m E_m)\varepsilon_{mu}$$

$$\sigma_c = \chi_1 \chi_2 V_f \sigma_f \big|_{\varepsilon_{mu}} + V_f \sigma_{mu} \qquad (3)$$

where

$\sigma_f \big|_{\varepsilon_{mu}}$ = fiber stress at matrix failure ($N \cdot m^{-2}$)

ε_{mu} = failure strain of matrix (dimensionless)
σ_{mu} = failure strength of matrix ($N \cdot m^{-2}$)

When the matrix fractures, the reinforcement suddenly has to bear the load transferred from the matrix. If this stress (σ_c) in equation (3) is greater than the stress the fibers can bear ($V_f \cdot \sigma_{fu}$), the membrane fails; otherwise, the fiber will continue to bear the load until the fibers break. Experimental evidence confirmed that cracks were observed in the matrix before the membrane fractured, i.e. $\varepsilon_{mu} < \varepsilon_{fu}$. From a waterproofing point of view, the membrane should be considered as failed at this point as it is no longer watertight. Therefore, the matrix properties governed the performance of the polyester reinforced membranes at –30°C.

The difference in the force-displacement curves of the samples reinforced with 180 $g \cdot m^{-2}$ non-woven polyester mats (Samples A2, B2 and C2) tested at –30°C further illustrated the important role of modified bitumen in membranes at low temperatures. The shapes of the force-displacement curves were markedly different because of different failure mechanisms. Samples A2 and B2 failed from the formation of a few deep cracks and sudden failure of the reinforcing mat. On the other hand, Samples C2 failed from one of the larger cracks which initiated from the multiple shallow cracks. Since the samples were reinforced by similar mats and at the same reinforcing density, it is evident that the modified bitumen governed failure mechanisms of the membrane at this temperature. Therefore, it is important to use modified bitumen with superior low temperature properties to achieve good low temperature performance in the membranes intended for cold weather use.

For the glass fiber reinforced membranes (A1, B1 and C1), the failure strain of the glass fiber reinforcement was lower than that of the bituminous matrix (i.e. $\varepsilon_{fu} < \varepsilon_{mu}$) at both test temperatures. It is evident from the Figure 2a and Table 3 that the elongation at break was about 4% at both 23°C and –30°C, as the breaking strain of the glass mat is not expected to be affected by test temperature. Therefore, the reinforcement controlled the failure of the membrane (i.e. $\varepsilon_c = \varepsilon_{fu}$). Equation (1) becomes:

$$\sigma_{cu} = (\chi_1 \chi_2 V_f \cdot E_f + V_m \cdot E_m)\, \varepsilon_{fu} \quad (4)$$

At 23°C, the modulus of the glass fiber reinforcement was much higher than the bituminous matrix ($E_f >> E_m$). Therefore, equation (4) is reduced to equation (2). The glass fibers bore most of the load exerted on the membrane and the membrane failed when the fiber broke.

However, as the test temperature dropped, the modulus of the modified bitumen increased and it began to carry an increasing proportion of the load exerted on the membrane ($V_m \cdot E_m \cdot \varepsilon_c$) so the strength of the membrane increased. This is similar to the polyester reinforcement. However, although the ultimate elongation of the matrix was greatly decreased when the temperature dropped to –30°C, the failure strain of the glass fiber mat was still lower (i.e. $\varepsilon_{fu} < \varepsilon_{mu}$). This is evident from the fact that the elongation at break of the membrane (strain at which the glass fiber mat broke) remained at 4% at both test temperatures. Therefore, equation (4) becomes:

$$\sigma_c = \chi_1 \chi_2 V_f \sigma_{fu} + V_m \sigma_m \Big|_{\varepsilon_{fu}} \quad (4a)$$

where

$\sigma_m \big|_{\varepsilon_{fu}}$ = matrix stress at fiber failure ($N \cdot m^{-2}$)

Experimental observations showed that the tensile elongation of the glass fiber reinforced membranes at both test temperatures were controlled mainly by the glass fiber reinforcement. However, the matrix carried a significant proportion of the load at low temperature, i.e. $V_m \sigma_m \big|_{\varepsilon_{fu}}$. To improve the strength of the membrane, one should focus on enhancing both the glass fiber reinforcement mat and the bituminous matrix. The reinforcement can be improved by increasing the fiber length (depending on the critical aspect ratio), rising the amount of reinforcement (V_f or a_f) or using a reinforcing cloth. Bitumen with high modulus at low temperature can also help to improve the strength of the membrane.

Low Temperature Flexibility and Glass Transition Temperature

The glass transition temperature (T_g) of the nine SBS-modified bituminous samples is shown in Table 5 and Figure 5. The lines in the figure were drawn to highlight the trends and do not have any statistical significance. The T_g of the control samples ranged from –48°C to –37°C. After six months of heat aging, the T_g increased to –43°C to –34°C. The change in T_g after six months of aging was less than +8°C. Although Sample C1 has the highest increase in T_g of +8°C, its T_g after six months of heat aging was –40°C, which remained low compared to the other samples.

Table 5 – *Effects of heat aging on the glass transition temperature.*

Sample ID	Glass Transition Temperature* (°C) after heat aging at 70°C for: 0 month	2 months	4 months	6 months
A1	-40	-42	-37	-34
B1	-40	-43	-40	-38
C1	-48	-50	-45	-40
A2	-40	-43	-39	-35
B2	-42	-45	-43	-40
C2	-37	-40	-39	-39
A3	-45	-46	-42	-38
B3	-41	-39	-39	-37
C3	-46	-49	-47	-43

* average of at least 2 specimens, measurement error is ±2°C

Bitumen has a glass transition temperature around –30°C [7]. The T_g (around –40°C) of the nine samples confirmed that SBS was effective in improving flexibility of the bitumen at low temperature. The glass transition temperature is affected by many factors such as the chemical composition and structure, molecular weight and intermolecular interactions of the polymer. The small change in T_g showed that the chemical properties of the SBS-modified bitumen in the samples did not seem to be significantly affected by heat aging. However, dynamic mechanical analysis might not pick up certain changes in molecular weight distribution in the modified bitumen. Gel Permeation Chromatography (GPC) analysis would provide such information and complement results from dynamic mechanical analysis. Analysis using GPC is recommended on the heat-aged samples in further study.

The cold flex temperature (T_{CF}) of the nine SBS-modified bituminous samples is shown in Table 6 and Figure 6. The T_{CF} of all samples increased after heat aging, with increases ranging from +3°C to +19°C. Most of the control samples had low T_{CF} (below –30°C) except for Samples A2 and C2. The T_{CF} of Sample C2 increased from about –10°C to -2°C (an average of –4°C in MD and +1°C in XD) – an increase of 8°C after six months of heat aging. The T_{CF} of Sample A2 increased from about -27°C to –10°C after six months of heat aging. This increase of 17°C was among the highest out of all samples tested.

Comparing Figures 5 and 6, T_{CF} and T_g showed similar trends of membrane flexibility with respect to heat aging although their actual values were quite different. T_{CF} measures membrane flexibility on a macroscopic level - the resistance to crack formation of the matrix under flexural load. T_g measures membrane flexibility on a microscopic level – the change in loss modulus under flexural load. Since T_{CF} and T_g measure different parameters for flexibility, they are affected by different factors in the membranes. Therefore, it was anticipated that the data might not show a direct correlation between T_{CF} and T_g.

It should also be noted that the T_{CF} of the two heat-aged samples, Samples C2 and A2, were high (–2°C and –10°C for Samples C2 and A2, respectively) but their T_g remained low (–39°C and –35°C for Samples C2 and A2, respectively). The cold flex temperature is affected by factors such as the thickness of the matrix, the chemical composition/structure of the modified bitumen, the type and amount of filler (e.g. limestone), and the bond strength between the filler and the bitumen. One possible explanation is that these two samples contained either an inappropriate amount of filler or filler which was not compatible with the bituminous matrix (this is consistent with observation that multiple shallow cracks were formed under tension at –30°C). The filler in the modified bitumen may reduce the T_{CF} but does not affect the T_g.

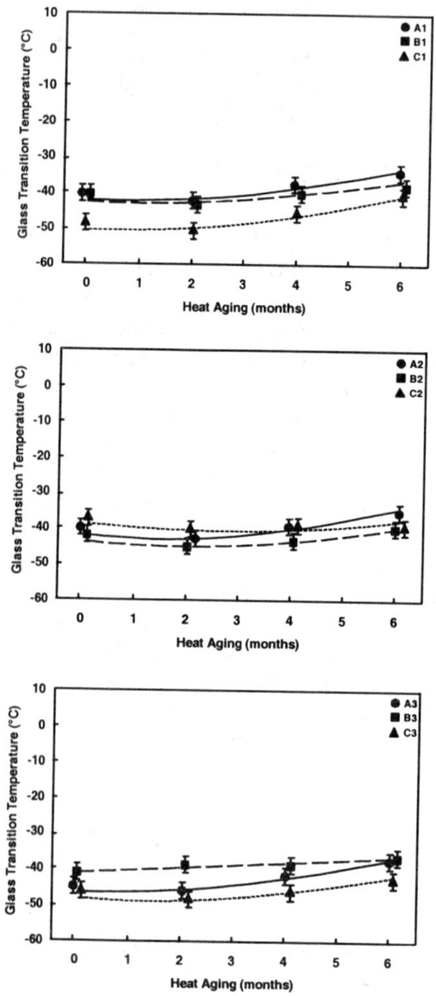

Figure 5 – *Effects of heat aging on the glass transition temperature of SBS-modified membranes.*

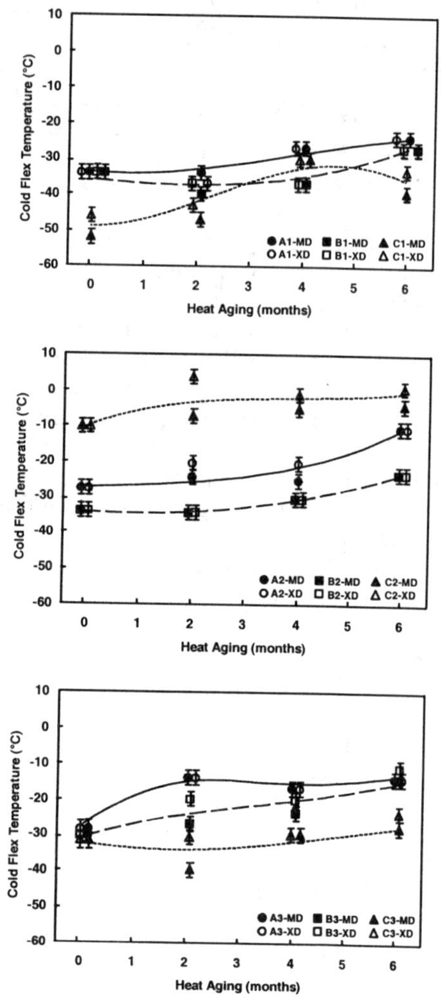

Figure 6 – *Effects of heat aging on the cold flex temperature of SBS-modified membranes.*

Table 6 – *Effects of heat aging on the cold flex temperature.*

Sample ID	Test Dir.	Cold Flex Temperature (°C) after heat aging at 70°C for: 0 month	2 months	4 months	6 months
A1	MD	-34	-34	-27	-24
	XD	-34	-37	-27	-24
B1	MD	-34	-40	-37	-27
	XD	-34	-37	-37	-27
C1	MD	-52	-47	-30	-39
	XD	-46	-43	-30	-33
A2	MD	-27	-24	-25	-10
	XD	-27	-20	-20	-10
B2	MD	-34	-34	-30	-23
	XD	-34	-34	-30	-23
C2	MD	-10	-7	-4	-4
	XD	-10	-4	-1	+1
A3	MD	-30	-14	-17	-14
	XD	-30	-14	-17	-14
B3	MD	-30	-27	-24	-14
	XD	-30	-20	-20	-11
C3	MD	-30	-40	-30	-27
	XD	-30	-30	-30	-24

Conclusions and Recommendations

1. Heat aging affected matrix-controlled properties such as elongation and strain energy but hardly affected the reinforcement-controlled properties such as breaking strength.
2. The heat aging effects on stress/strain was more pronounced at –30°C and in membranes containing less reinforcement.
3. The glass fiber mat controlled the tensile strength of the membrane at 23°C but the matrix contributed a significant portion to the strength of the membrane at –30°C. The glass fiber mat controlled the tensile elongation of the membrane at both temperatures.
4. The polyester reinforcement controlled the tensile properties of the membrane at 23°C but the matrix properties governed the membrane performance at –30°C.
5. The low glass transition temperature (around –40°C) of the membranes before and after heat aging indicated that the SBS-modified membranes were suitable for cold weather application and not affected significantly by heat aging.
6. The cold flex temperature of the membranes increased by 3°C to 19°C after six months of heat aging. The samples that had high cold flex temperature also experienced high reduction in elongation after heat aging. This is a direct indicator of the changes in the bitumen matrix during aging.
7. Since heat aging affects matrix-controlled properties, which governs the membrane performance at low temperature, it is important to use high quality modified bitumen in order to manufacture durable membrane for cold weather application.

Acknowledgement

The authors gratefully acknowledge the assistance of H. Shen, J. Irwin and A. Delgado in conducting some of the laboratory testing, and the financial support of Soprema Inc.

References

[1] National Research Council Canada, *National Building Code of Canada*, Appendix C, 1995.

[2] May, J. O., "Temperature profiles of different roof waterproofing systems subjected to natural exposure conditions," *Proceedings of the 2nd International Symposium on Roofing Technology*, pp. 80-85.

[3] Duchesne, C., "Durability of the SBS-modified bituminous double-layer system: correlation between performances after artificial and natural aging," *Proceedings the 3rd International Symposium on Roofing Technology,* pp. 222-226.

[4] Braun, E. and Hendriks, N., "Comparison of six different accelerated weathering tests on modified bitumen products," *Roofs and Roofing*, Chapter 26, May, J.O. Ed., Ellis Horwood Ltd, England, 1988, pp. 315.

[5] Puterman, M. and Marton, M. "Evaluation of changes in roofing materials as a result of long-term exposure," *Proceedings the 4th International Symposium on Roofing Technology,* pp. 236-241.

[6] Baxter, R. and Kearney, T. "Weathering characteristics of polymer modified asphalt roofing membrane," *Proceedings the 3rd International Symposium on Roofing Technology,* pp. 453-458.

[7] Rodriguez, I., Dutt, O., Paroli, R.M. and Mailvaganam, N.P., "Effect of heat-aging on thermal and mechanical properties of APP- and SBS-modified bituminous roofing membranes," *Materials and Structures*, Vol. 26, No. 160, pp. 355-361.

[8] CGSB 37.56-M (9th draft), "Modified Bituminous Membranes for Roofing," Canadian General Standards Board, January 1997.

[9] Piggott, M.R., *Load Bearing Fibre Composites*, 2nd edition, MERP Enhanced Composites Inc., Toronto, 2001, p96, p169, pp. 157.

[10] Kardos, J.L., "Short-Fiber-Reinforced Polymeric Composites, Structure-Property Relations," *International Encyclopedia of Composites*, Vol. 5, ed. Lee, S.M., VCH Publishers Inc., 1991, pp. 133.

Ming L. Shiao,[1] David A. Nester,[2] and Louis A. Terrenzio[2]

On the Kinetics of Thermal Loads for Accelerated Aging

REFERENCE: Shiao, M. L., Nester, D. A., and Terrenzio, L. A., **"On the Kinetics of Thermal Loads for Accelerated Aging,"** *Roofing Research and Standards Development, 5th Volume, ASTM STP 1451,* W. J. Rossiter and T. J. Wallace, Eds., ASTM International, West Conshohocken, PA, 2003.

ABSTRACT: A mathematical model based upon the theory of total cumulative damage has been developed to describe the kinetics of thermal aging in roofing material subjected to changing temperatures and durations. The new model employs a degradation rate function that is derived from a chemical model using a diffusion theory coupled with Arrhenius rate function to describe the temperature dependency of asphalt chemical composition under transient conditions. Such a model enables the construction of master curves in asphalt composition as aged at various temperatures, and may be used to predict the change in property due to heat aging using known roof temperature profile and materials parameters obtained from lab accelerated testing. Examples of the model are demonstrated by using the real-time roof temperature data collected from test houses located in a cold climate of Minnesota and in a hot climate of Florida. The calculations suggest that the high thermal history in a hot climate has a strong effect in accelerating aging of asphalt, particularly during the winter months. Furthermore, the effect of attic ventilation is found to reduce the uneven distribution of total cumulative damage across the roof deck due to unbalanced thermal history found in an unventilated deck. The potential application of the model and its advantage in predicting field properties by simulation are also discussed.

Keywords: aging, kinetics, asphalt, shingle, cumulative damage, thermal load, ventilation, accelerated test, temperature effect

Introduction

Asphalt shingles have been successfully used as roof coverings for more than a century with proven durability and longevity [*1-2*]. Such success is not only a product of continuous research as the roofing industry strives for better products and manufacturing excellency [*1,3*], but is also accumulated from many costly experiences of premature failures [*4-6*]. Hence, the durability of roofing materials and their corresponding life predictions in field conditions have long been a subject of extensive study [*3,7-12*]. Furthermore, with new advances in raw materials for roofing applications and increasing complexity of shingle designs, it is now more desirable to have the capability to design and predict field properties ahead of time in order to enhance product performance and

[1] Principal Research Engineer, CertainTeed Corp., Exterior Products Group, 1400 Union Meeting Road, Blue Bell, PA 19422

[2] Principal Chemist and Director of Technical Service, respectively, CertainTeed Corp., Exterior Products Group, 1400 Union Meeting Road, Blue Bell, PA 19422

prevent costly premature failures. As a result, many accelerated test methods and procedures have been developed in the laboratory to evaluate the durability of roofing materials [*13*]. However, their relationships to life predictions under the field conditions are often difficult, since most lab accelerated aging or weathering tests are often limited by relatively short exposure to controlled environments in steady state, whereas in field conditions the roofing materials are constantly interacting with a varying environmental loadings [*11,12*]. In many cases, the so-called "Elephant Tests" are imposed on the materials by using exceptionally high level of loadings to accelerate the failure. Although proper lab testing by accelerated aging has been successful in predicting the performance of roofing materials, the lack of theoretical models in linking lab testing under steady conditions to the real-life, transient conditions is yet to be addressed.

On the other hand, recent advances in predicting service life using empirical survey data or reliability-based analysis from performance inspection data have been proposed [*9,10*]. Although these predictions may be useful for end users such as building owners to assess and manage the roof systems, they are limited in their ability to predict the performance of roofing products constructed with new or improved materials and designs. Furthermore, predictions based upon such an approach can sometimes lead to erroneous conclusions regarding the environmental loading of outdoor exposures, e.g., thermal loading or UV radiation, since these models are often the result of analysis of financial or performance inspection data and are not necessarily based upon true physical or chemical degradation models. Hence, new models may need to be developed, particularly from the materials point of view, in order that the aging behavior of roofing materials may be properly predicted. Recent advances in modeling stochastic process for application in survival analysis under highly variable environments have been proposed [*14*]. These theories may be suitable for linking varying outdoor exposures to the property change of a particular roofing material by adopting a proper degradation rate model, which can be obtained from laboratory testing under steady conditions. Thus, by using proper real-time data of desirable environmental loading, or by simulating the environmental loading using weather station data [*15*], one may be able to virtually construct from such theory the aging or weathering history of the particular roofing material. These calculations can then be used to assist in better prediction of the future field property of roofing materials to improve product design or materials selections.

In this paper, a degradation model due to varying thermal loads is developed for asphaltic roofing materials based upon the theory of total cumulative damage and a chemical degradation rate function obtained from a heat aging study of asphalt. The model is then applied to two scenarios of asphalt shingles exposed in a cold and hot climates, both with and without attic ventilation, by using real-time roof temperature data to estimate the rate of heat aging. The results are then discussed in terms of the accumulated thermal history and the thermal distribution of the roof, as well as the effect of ventilation to shed light on the potential implications of such a model and its advantage in predicting future roofing material performance by simulations.

General Theory for Degradation Rate in Outdoor Environments

To obtain the kinetics of degradation as a function of varying environmental parameters, the theory of total cumulative damage is used [*16-18*]. Such theory has been developed to study the degradation-based life data under changing stress levels and an excellent survey of the related models has been given by Saunders [*19*]. The model of cumulative damage was also used recently by Bauer et al. [*20*] and by Meeker et al. [*21*] to study the service life prediction for materials exposed to highly variable outdoor exposures. The general form for the total cumulative degradation at time t can be given by [22]

$$D(t;\Psi[0,t],\theta) = D_0 + \int_0^t h(\tau;\Psi(\tau),\theta)d\tau \tag{1}$$

where D_0 denotes the initial degradation at time 0; $h(\tau;\Psi[\tau], \theta)$ represents the degradation rate function; and $\Psi[\tau]$ is the state of the outdoor environmental factors in the time interval $[0,t]$. θ is a vector of parameters derived from a physical/chemical model that relates degradation rate to the outdoor variables. Equation (1) has been applied recently by Chan and Meeker [22] to study the degradation of a solar reflector material with degradation rate function obtained by Jorgensen et al. [*23*] as a function of UV radiation $U(t)$, temperature T, and relative humidity R given as

$$h(t;\Psi(t),\theta) = AU(t)T(t)^{-B}e^{-\varepsilon/T(t)}e^{R(t)(C+D/T(t))} \tag{2}$$

where A, B, C, and D are constants. By using the daily weather data obtained from the National Oceanic and Atmospheric Administration (NOAA), they were able to predict the daily degradation on a solar reflector as occurred in Boulder, CO and in Fort Peck, MT by using time series model to simulate the incremental degradation accumulated within each 24-hour period.

To estimate the total accumulative damage in asphaltic roofing materials due to outdoor exposures such as heat, a degradation rate function based upon useful physical or chemical model needs to be first developed. Such a model should provide the link between the property change in asphalt and the corresponding temperature exposures. However, survey on the current literature of roofing materials indicates that these types of relationships are yet to be established. In the following, a chemical degradation model for asphalt due to thermal loads will be examined by studying the change in asphalt chemical composition as a function of aging time and aging temperature. Subsequently, the related degradation rate function and the general form for total cumulative damage in asphalt subjected to varying temperatures will be established by linking the change in chemical composition to the total level of thermal loads received. The resulting degradation model for asphalt material will then be used to estimate the total cumulative damage in cold and hot climates using asphalt shingle temperature data obtained from the test houses located in Cloquet, Minnesota and in Gainesville, Florida. Details of the experimentation will be discussed in the following sections.

Chemical Model of Asphalt Degradation Due to Heat Aging

In the case of thermal loading, our previous work [*24*] has shown that a diffusion model can be used to describe the aging behavior of asphalt and shingles under lab accelerated aging and outdoor aging conditions. The model describes the property or performance change in asphalt or shingles due to the heat aging as

$$\frac{P_t(t,T)}{P_0} = 1 + \alpha e^{-\beta(T)/\sqrt{t}}, \quad t > 0 \tag{3}$$

where $P_t(t,T)$ is the physical or chemical property at aging time t and P_0 denotes the initial property at t=0. α and $\beta(T)$ are two parameters that define the aging process in terms of its final stage and the rate of reaching such stage through diffusivity of oxidation, as well as through the phase separation of low molecular weight oils in asphalt. In previous paper [*24*], the model shown in Equation (3) has been successfully applied to fit the aging data obtained by various accelerated aging test methods, as well as data from outdoor exposures in Minneapolis, Philadelphia, and Houston. However, the exact form of the temperature dependence in Equation (3) was not established in the paper.

In this paper, the temperature dependence of Equation (3) is established by examining the chemical composition of asphalt aged in a dark oven at various temperatures. Coating grade asphalt from a Canadian source was prepared and pressed into thin films that were placed in forced-air dark ovens maintained at 40°C, 60°C, 70°C, and 80°C for heat aging up to a year. Small amounts of sample were then periodically collected from the thin film samples and were tested for fractionation analysis by Corbett procedure, as well as the analysis for molecular weight (MW) distribution using gel permeation chromatography. The details of the experimentation have been discussed in Ref *24* and therefore will not be repeated in here.

The change in asphalt chemical composition due to heat aging, e.g., the weight % change in asphaltene content, in naphthene (N.) aromatics content, and the increase in the averaged MW, are tabulated in Table 1 as a function of aging time and aging temperatures. The data are also normalized in Fig. 1 through Fig. 3 with respect to their initial values as a function of aging time and aging temperatures. These results clearly show the strong effects of temperature on asphalt chemical composition by thermal oxidation and are also consistent with those results typically found in the literature [*25*].

Prior to the data analysis using the diffusion model of Equation (3), the results in Table 1 were first examined by using models such as Arrhenius equation, Eyring rate equation, exponential model, and power law model. However, none of these models yield satisfactory fitting to the overall data set. For example, the data analysis by the Arrhenius equation reveals that the activation energy is not a constant and is strongly affected by the aging time. Furthermore, the form of such time dependence is also found to be inconsistent among the three chemical properties examined. For the power law model, even though it is found to be more consistent in describing the overall data set, however, the regression coefficients are found to be relatively low in many cases and its physical meaning is also not as clear. On the other hand, the analysis of the overall data set in Table 1 using the diffusion model of Equation (3) has consistently yielded good results among the three chemical properties affected by heat aging. By plotting $ln(P_t/P_0-1)$ versus $1/\sqrt{t}$, the two parameters of α and β can be determined through a least-square linear regressions. An example of such regression is shown in Fig. 4 for the case of asphaltene content. The results of such regressions for the two parameters α, β and their corresponding regression coefficients are shown in Table 2. As one can see, the

Table 1 – *Chemical analysis of asphalt by dark oven aging at various temperatures.*

Aging Time, hours	Chemical Property	Temperature			
		40°C	60°C	70°C	80°C
0	Avg. MW	2115	2115	2115	2115
	% Asphaltene	31.01	31.01	31.01	31.01
	% N. Aromatics	30.30	30.30	30.30	30.30
336	Avg. MW	2246	2453	2762	2960
	% Asphaltene	35.30	38.86	41.17	42.85
	% N. Aromatics	28.54	24.31	23.11	22.69
672	Avg. MW	2298	2554	2907	3249
	% Asphaltene	36.02	40.05	44.54	45.76
	% N. Aromatics	26.68	24.50	21.43	20.34
1344	Avg. MW	2347	2709	3214	3728
	% Asphaltene	36.73	43.47	47.05	49.78
	% N. Aromatics	26.65	22.07	18.79	17.84
2016	Avg. MW	2394	2830	3382	- -
	% Asphaltene	37.46	44.82	49.42	54.46
	% N. Aromatics	26.21	20.44	17.73	15.36
4392	% Asphaltene	39.49	46.67	51.87	58.79
	% N. Aromatics	23.20	18.79	16.62	13.64
8760	% Asphaltene	41.31	- -	- -	- -
	% N. Aromatics	21.01	- -	- -	- -

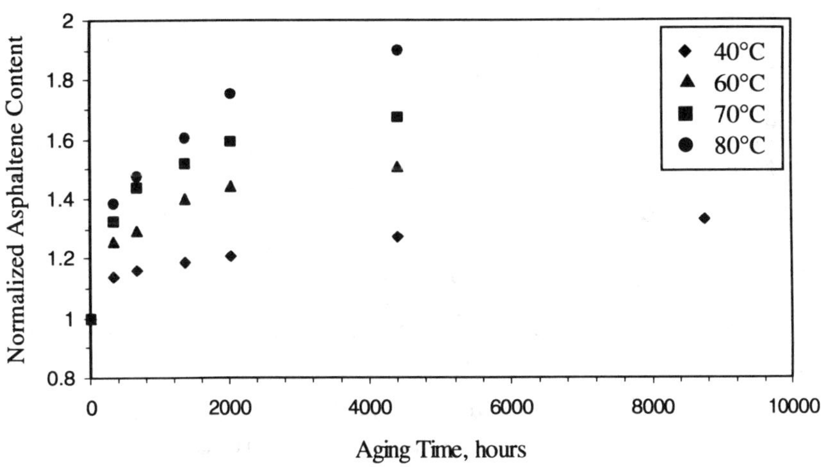

Figure 1 – *The temperature effects of aging on the % asphaltene contents.*

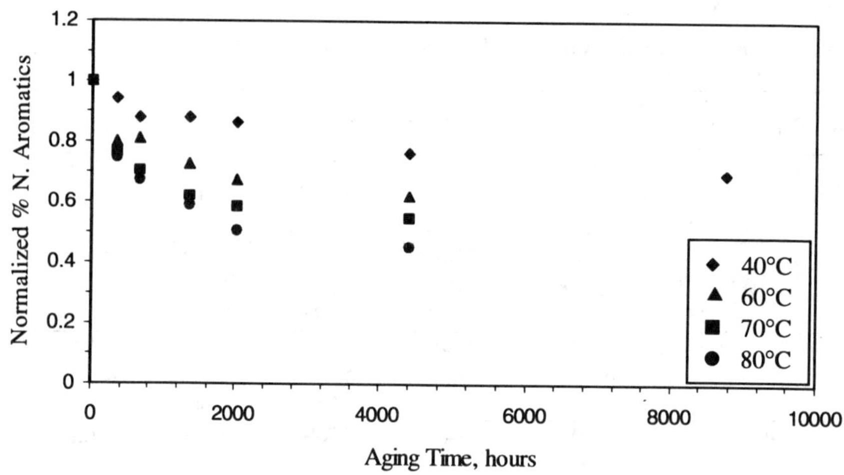

Figure 2 – *The temperature effects of aging on naphthene aromatics contents.*

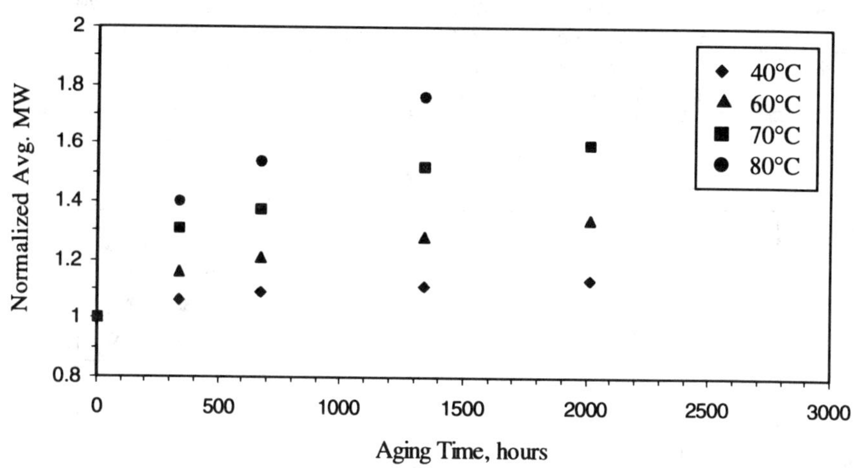

Figure 3 – *The normalized averaged MW as a function of aging time and temperature.*

regressions are in very good agreement with the diffusion model of Equation (3), except for the case of % naphthene aromatics aged at 40°C, where the data appears to be much noisier. Note that for all of the three chemical properties examined, the parameter β is found to be relatively constant with respect to aging temperature, whereas the parameter α is found to be strongly influenced by the temperature according to an Arrhenius form.

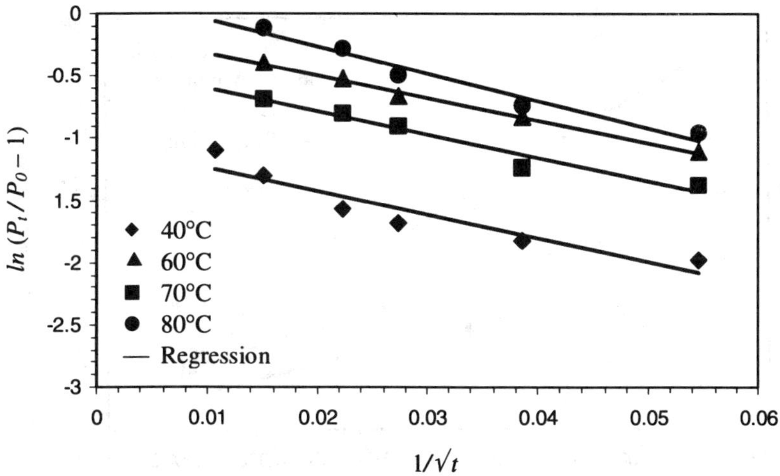

Figure 4 – *Example of data analysis using Eq. (3): % asphaltene due to heat aging.*

Table 2 – *The regression of heat aging data in Table (1) using Eq. (3).*

Temperature	% Asphaltene			% N. Aromatics			Avg. MW		
	α	β	R^2	α	β	R^2	α	β	R^2
40°C	0.35	19.02	0.89	-0.29	26.10	0.65	0.21	22.69	0.99
60°C	0.66	18.45	0.96	-0.46	17.79	0.83	0.53	22.78	0.97
70°C	0.88	18.14	0.99	-0.59	17.10	0.99	0.92	20.92	0.96
80°C	1.18	21.82	0.96	-0.74	20.33	0.99	1.40	23.35	0.98
Average		19.36			20.33			22.43	

This can be seen by plotting $ln\,|\alpha|$ versus the inverse of aging temperature $1/T$ in °K as shown in Fig. 5, where the values of $ln|\alpha|$ is found to vary linearly with the inverse of aging temperature T for all of the three chemical properties. This allows the temperature dependence of parameter α to be written as

$$\alpha = n \cdot e^{-E/kT} \tag{4}$$

where n is a constant and E is the activation energy. k is the Boltzmann constant. The values of the activation energy E and the constant n are listed in Table 3 along with their coefficients of regression. Again, the regression coefficients all show excellent agreements. The values of the activation energy E are found to be about 0.23~0.46 eV, which are also in excellent agreement with the activation energy observed in the carbonyl formation during oxidation of asphalt [26]. As a result, the temperature dependence of

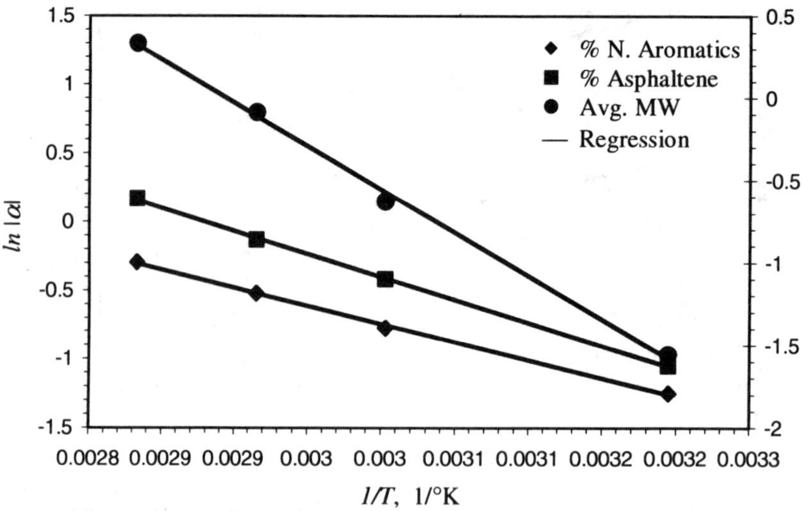

Figure 5 – *The relationship of parameter α and the aging temperature T.*

Table 3 – *The regression result for α versus aging temperature.*

Property	*n*	*E, eV*	R^2
% Asphaltene	15081.14	0.2877	1.000
% N. Aromatics	1279.08	0.2267	1.000
Averaged MW	3.9373×10^6	0.4518	0.999

Equation (3) now can be expressed as

$$\frac{P_t(t,T)}{P_0} = 1 + n' \cdot e^{-E/kT - \beta/\sqrt{t}} \qquad (5)$$

where $n'=n$ except for the case of % naphthene aromatics, in which $n'=-n$. The strong temperature effect in Equation (5) is somewhat unexpected, since it suggests that the heat aging in asphalt is accelerated towards an end level that is dominated by temperature. Although the exact physical interpretation of such dependency is unclear at this point, however, it is suspected that the heat aging is greatly accelerated beyond simple Fickian behavior as aging time and temperature increase, where the change in molecular structure and phase separation of low-end MW substance can significantly affect the process of oxidation. Nevertheless, the simplicity of equation (5) enables one to shift the aged property at different aging temperatures along the time scale by calculating the equivalent aging time t_f in the referenced aging temperature T_f as

$$t_f = 1 \Big/ \left[\frac{E}{\beta k}\left(\frac{1}{T} - \frac{1}{T_f}\right) + \frac{1}{\sqrt{t}}\right]^2 \tag{6}$$

This allows one to generate a master curve using all the data aged at various temperatures and examples of such master curves are shown in Fig. 6 and Fig. 7 using the reference temperature of 80°C. As one can see, the chemical composition data obtained by heat aging at various aging temperature can now all be linked together along the time axis, which also shows an excellent agreement with the theoretical predictions based on Equation (5).

General Model of Total Accumulative Damage for Asphalt Due to Heat

In the case of varying thermal loading, the change in property can then be derived from Equation (5) by assuming that the effect of heat aging on property is additive and that the Boltzmann superposition principle is applicable. Hence, one can rewrite the Equation (5) as

$$\Delta P(t, T(t)) = P_t - P_0 = P_0 \cdot \int_0^t \frac{P(t-\tau, T(\tau))}{T(\tau)} \frac{dT(\tau)}{d\tau} d\tau \tag{7}$$

or

$$\Delta P(t, T(t)) = n' P_0 \int_0^t \frac{e^{-E/kT(\tau) - \beta/\sqrt{t-\tau}}}{T(\tau)} \frac{dT(\tau)}{d\tau} d\tau \tag{8}$$

Hence, the total cumulative degradation at time t due to varying heat aging can be derived by the notion that the amount of property change may be directly related to the total amount of exposure or damage due to heat aging. Thus,

$$D(t; \Psi[0,t], \theta) \propto \left|\Delta P(t, T(t))/P_0\right| = A\ n \int_0^t \frac{e^{-E/kT(\tau) - \beta/\sqrt{t-\tau}}}{T(\tau)} \frac{dT(\tau)}{d\tau} d\tau \tag{9}$$

where A is a proportional constant and D_0 from Equation (1) is omitted. Here the vector $\Psi[0,t]$ for the outdoor environmental factor of temperature is given by

$$\Psi(t) = \frac{n \cdot e^{-E/kT(t)}}{T(t)} \frac{dT(t)}{dt} \tag{10}$$

and the parameter vector θ includes those regression constants of β, n, E, and A. The

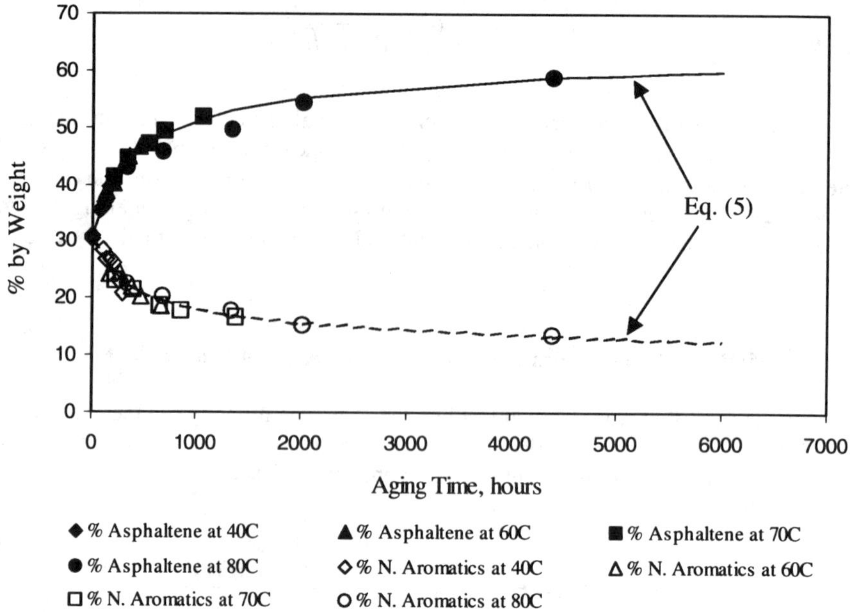

Figure 6 – *The master curves for % asphaltene and % N. aromatics.*

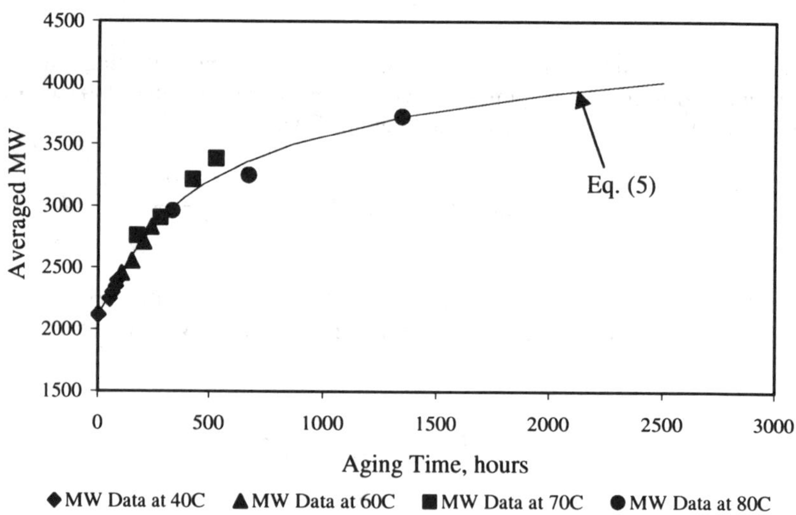

Figure 7 – *The master curve for averaged MW.*

degradation rate function can therefore be written as

$$h(t;\Psi(t),\theta)=n\cdot\frac{e^{-E/kT(t)-\beta/\sqrt{t}}}{T(t)}\frac{dT(t)}{dt} \tag{11}$$

Thus, if the temperature profile as a function of time is known, Equation (9) can then be used to estimate the total amount of damage in asphalt due to absorption of solar heat or heat build-ups from its surroundings. Equation (9) can also be written in its incremental form as

$$D(t;\Psi[0,t],\theta)=An\sum_{\tau=0}^{t}\frac{e^{-E/kT(\tau)-\beta/\sqrt{t-\tau}}}{T(\tau)}\Delta T(\tau) \tag{12}$$

The proper choice of the set of vector θ will depend on the interpretation of chemical composition in relationship to the failure mechanisms that are important under one's consideration. For example, the amount of increase in asphaltene content in asphalt has been shown to relate to the increase of softening point, hardness, viscosity, and to the decrease of asphalt durability [27]. Thus, for those who consider the hardening of asphalt is largely responsible for the failure of a particular roofing system, the use of the vector set θ derived from the % change in asphaltene content due to heat aging will then deem to be appropriate. In the following, the concept of total cumulative damage as shown in Equation (12) will be demonstrated by using real-time roof temperature data obtained from test houses located in extreme cold and hot climates. The θ vector set of asphaltene content will be used as an example to estimate the total cumulative damage due to varying heat loadings.

Calculation of Total Cumulative Damage Due to Heat Using Real-Time Roof Data

To examine the effects of varying outdoor temperature exposure on asphalt shingles in a typical roof assembly, the amount of total cumulative damage is calculated using the roof sheathing temperature data collected from the test houses located in the northern cold climate at Cloquet, Minnesota and in the southern hot climate at Gainesville, Florida. Both test houses were constructed by similar construction practice and by the same configurations that allow paired test bays for studying the effects of attic types, ventilations, wall systems, and various insulation configurations. The details of the configuration and instrumentation of the test houses have been discussed by Goldberg et al. [28] and by Porter et al. [29]. For current analysis, the temperature data obtained from the test bays of flat ceiling with and without ridge ventilation will be used. The temperature data were collected from thermal couples placed underneath a 25 year warranted, 3-tab shingle on a south-facing deck. The roof has been exposed for four years when the temperature data was collected. Two thermal couple positions, one near the soffit edge and the other near the roof ridge along the centerline of roof deck, were used to collect the temperature data. For the Test House in Florida, the soffit sensor is located at 2.5 ft from the lower edge and the ridge sensor is at 2.7 ft from the top edge of the deck. For the Minnesota Test House, the soffit sensor location is at 9 in. from the soffit edge and the ridge sensor is located at 11 in. from the ridge along the centerline of the deck. The work of data collection was carried out between July 2001 and August 2002 through the collaboration with the University of Minnesota and the University of Florida. The roof temperature data were sampled at an interval approximately every 15 minutes in the Florida test house and about every 23 minutes in the Minnesota test house. This represents a very detailed roof temperature profile with data points in excess of 20,000 to 30,000 per each sensor channel. An example of such a profile for the temperature data collected from the soffit sensor in the vented test bay at the Florida Test House is shown in Fig. 8.

Equation (12) was used to calculate the total amount of cumulative damage due to temperature by omitting the proportional constant A. The calculation was done in summation to a monthly basis from August 2001 to the end of July 2002, so that the seasonal effects on temperature change can be examined. This is equivalent to the difference between asphalt shingles that are freshly exposed at both climates for a particular month only. The resulting total cumulative damages for the two Test House locations are listed in Table 4 and they are also charted in Fig. 9 to Fig 10 for comparison purpose. As one can see, the % total cumulative damage as calculated from the roof temperature data shows that the asphalt shingle exposed in Florida results in significantly higher cumulative damage due to the higher amount of heat loading. This results in the estimated chemical change of asphalt to be about 4 times faster when compared to the same exposure period in the cold climate of Minnesota. It is also interesting to note that in winter months the damage due to heat aging in Florida can be as much as 10 times faster than in Minnesota, whereas in summer months the difference is only about 2 to 3 times faster in the Florida climate. For the effects of attic ventilation, it is noted that in both test house locations the asphalt shingle in the vented attic is found to have more uniform cumulative damage between the soffit and ridge locations. On the other hand, in the unvented attic a significant difference in the amount of total cumulative damage is observed between the soffit and the ridge locations. This result has the important implication that the thermal history of asphalt shingles exposed in a unvented attic can have higher degree of difference between the ridge and soffit locations due to higher degree of uneven heat distribution along the roof deck. This can result not only in a greater degree of difference at the rate of aging, but also creates higher amount of thermal stresses between top and bottom portion of the roof. Such difference is especially strong in the hot climates of Florida, where the asphalt shingles in the ridge location are found to have the highest degree of cumulative damage. For the climate at the Minnesota Test House, the difference between the soffit and ridge locations in the unvented attic is still noticeable but not as significant.

Conclusions

A mathematical model based upon the theory of total cumulative damage has been developed to estimate the rate of degradation in asphaltic roofing materials that are subjected to changing environmental loading such as heat. This model is derived from a degradation rate function that is stemmed from a chemical degradation study based upon the change in asphalt chemical composition as a function of aging time and aging temperature. The change in asphalt chemical composition is found to be described by the diffusion model developed by the authors [*24*], with a modified form to accurately predict the temperature dependency. The proposed new chemical degradation function also allows the construction of master curves of chemical composition due to heat aging by shifting the data at various aging temperatures along the time axis.

The derived model of total cumulative damage has been applied to estimate the amount of shingle damage due to thermal loading at extreme cold and hot climates by using the roof temperature data from Test Houses at Cloquet, MN and Gainesville, FL. The calculations show that the total amount of heat aging in a typical year at Florida can be four times faster than at Minnesota during the same period. Also, the effects of attic ventilation are found to produce more uniform heat aging along the roof deck, whereas in unvented attic significant difference in the amount of total cumulative damage is found between the top and bottom portion of the roof deck.

Acknowledgments

The authors would like to acknowledge the College of Architecture and Landscape Architecture at the University of Minnesota, Minneapolis, and the College of Architecture at the University of Florida, Gainesville, for providing the roof temperature data. Helpful data arrangement by Dr. Louise F. Goldberg and comments on the data collection of roof temperatures from David Ober are greatly appreciated. The authors are also thankful for John Harrison who carried out the Corbett analysis for asphalt chemical composition.

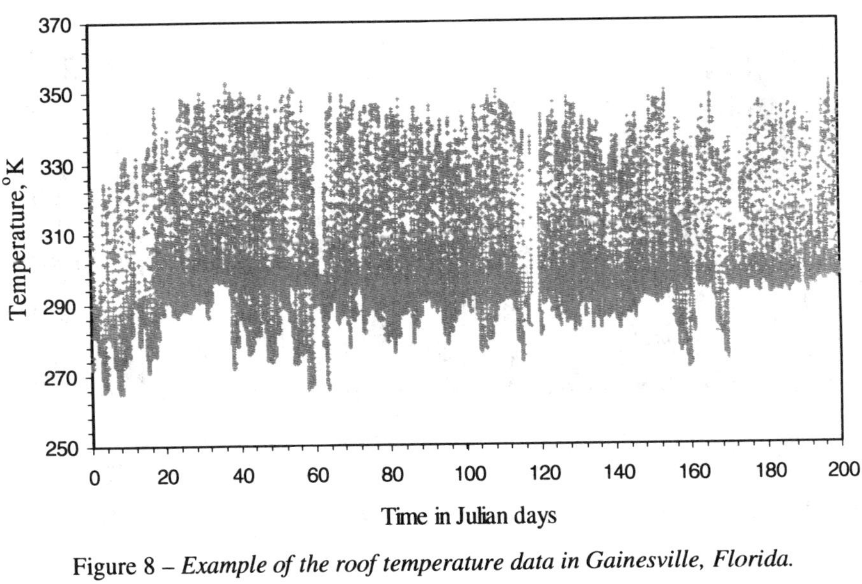

Figure 8 – *Example of the roof temperature data in Gainesville, Florida.*

Table 4 – *The calculated monthly % total cumulative damage due to heat aging.*

Month	Gainesville, Florida				Cloquet, Minnesota			
	Vented Attic		Unvented Attic		Vented Attic		Unvented Attic	
	Soffit*	Ridge*	Soffit	Ridge	Soffit	Ridge	Soffit	Ridge
Aug./2002	19.95	21.26	18.22	21.64	7.30	7.85	6.79	8.89
Spet.	19.04	20.09	16.73	19.38	5.38	5.36	4.67	6.06
Oct.	11.92	12.35	12.06	15.08	4.02	3.67	3.40	4.02
Nov.	10.64	10.50	9.18	10.36	2.29	2.04	1.92	2.12
Dec.	8.78	8.47	8.40	9.31	0.94	0.71	0.74	0.71
Jan./2003	7.39	7.14	8.44	8.93	1.39	0.77	1.10	0.73
Feb.	5.84	5.79	5.57	6.22	1.77	1.63	1.52	1.70
March	12.39	12.71	12.41	13.82	2.58	0.71	2.12	0.47
April	15.29	16.24	14.02	16.62	4.10	4.34	3.70	4.62
May	12.42	13.16	12.04	13.97	4.69	5.20	4.34	5.66
June	14.60	16.18	12.68	15.76	6.22	7.06	5.74	7.86
July	18.00	19.84	16.62	19.72	7.02	7.72	6.57	8.72
Total	156.25	163.74	146.37	170.80	47.70	47.07	42.63	51.56
Average	13.02	13.65	12.20	14.23	3.98	3.92	3.55	4.30

*: The soffit and ridge locations are the thermal couple locations for collecting roof temperature data, which represent the upper and lower portion of the temperature profile along the roof deck.

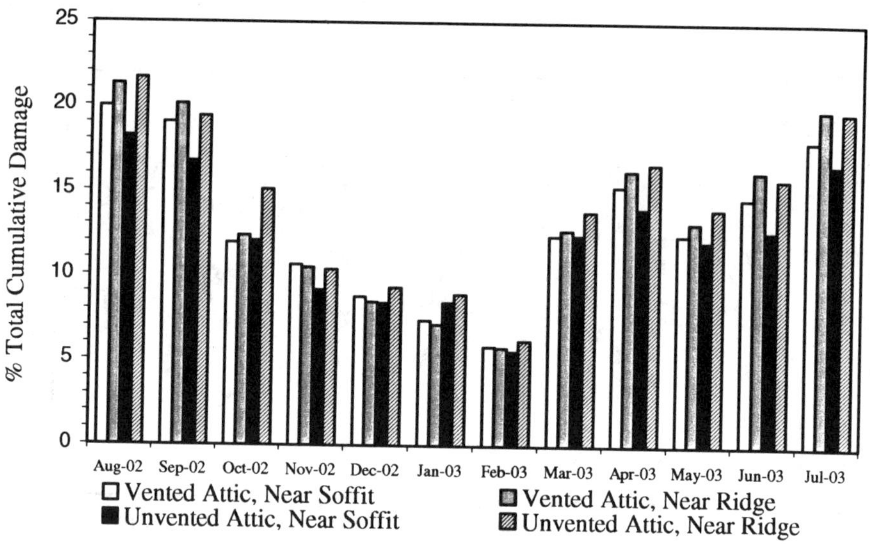

Figure 9 – *The calculated % total cumulative damage in the hot climate of Florida.*

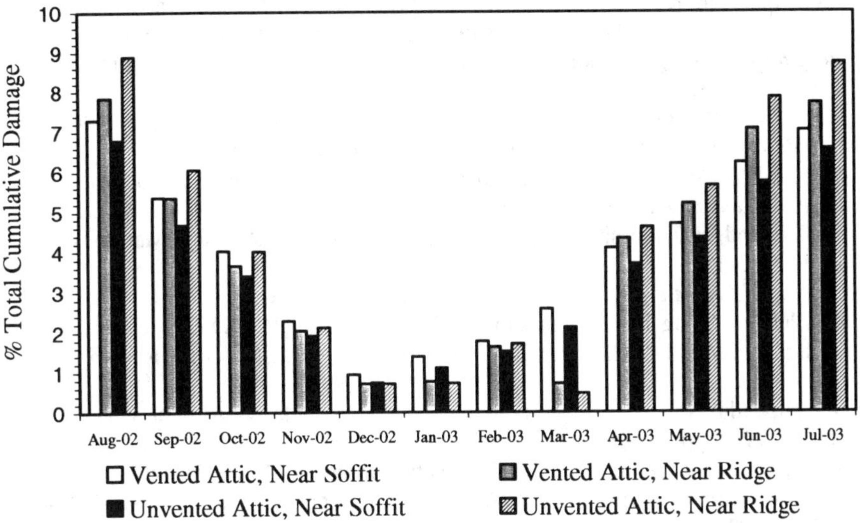

Figure 10 – *The calculated % total cumulative damage in the cold climate of Minnesota.*

References

[1] Noone, M. J. and Blanchard, W. K., "Asphalt Shingles-A Century of Success and Improvement," *Proceedings of the 10th Conference on Roofing Technology*, National Roofing Contractors Association, Rosemont, IL, 1993, pp. 23-33.

[2] Adler, D. L. and Cash, C. G., "Asphalt Shingles: a Primer," *Construction Specifier*. Nov. 1995, pp. 24-33.

[3] Cullen, W. C., "Research and Performance Experience of Asphalt Shingles," *Proceedings of the 10th Conference on Roofing Technology*, National Roofing Contractors Association, Rosemont, IL, 1993, pp. 6-12.

[4] Ribble, R., Olson R., Summers D., and Goodman J., "From Generation to Generation: Issues and Problems Facing the Steep-Slope Roofing Industry," *ibid.*, pp. 1-5.

[5] Cullen, W. C., "Problems Prompt a Quest for High Quality Standards," *Roofing Research and Standards Development: 3rd Volume, ASTM STP 1224*, T. J. Wallace and W. J. Rossiter, Eds., American Society for Testing and Materials, 1995, pp. 3-14.

[6] Cullen, W. C., "The Evolution of Asphalt Shingles: Survival or the Fittest?," *Professional Roofing*, June 1992, pp. 4-8.

[7] Akoz, F. and Akman, M. S., "Service Life Estimation for Mutli-ply Flat Roof Membranes, " *Durability of Building Materials and Components 5*, 1990, pp. 353-358.

[8] Cash, C. G., "Durability of Bituminous Built-Up Roofing Membranes," *Durability of Building Materials and Components, ASTM STP 691*, Sereda P .J. and Litvan G. G., Eds., 1980, pp. 741-754.

[9] Kyle, B. R. and Kalinger P., "Service Life Prediction of Roof Systems by Reliability-Based Analysis," *Proceedings of the 4th International Symposium on Roofing Technology*, National Roofing Contractors Association, Rosemont, IL, 1997, pp. 50-57.

[10] Cash, C. G., "The Relative Durability of Low-Slope Roofing," *ibid.*, pp. 119-124.

[11] Lounis Z., Lacasse, M. A., Vanier, D. J., and Kyle, B. R., "Towards Standardization of Service Life Prediction of Roofing Membranes," *Roofing Research and Standards Development: Fourth Volume, ASTM STP 1349*, T.J. Wallace and W. J. Rossiter, Eds., American Society for Testing and Materials, 1999, pp. 3-18.

[12] Nelson, W., *Accelerated Testing*, John Wiley & Sons, New York, 1990.

[13] Accelerated Aging Test Program for Asphalt Shingles, ARMA Research Committee, May 1995, Asphalt Roofing Manufacturers Association, Rockville, MD.

[14] Singpurwalla, N. D., "Survival in Dynamic Environments," *Statistical Science*, vol. 10, 1995, pp. 86-103.

[15] Sailor, D., Akbari H., and Martien P., "On the Ability to Model Near-Surface Urban Air Temperatures," Lawrence Berkeley National Laboratory Report LBL-31406, 1992, Berkeley, CA.

[16] Chapter 11 in Reference [12].

[17] Tobias, P. A. and Trindade, D. C. *Applied Reliability*, 2nd Edt., Van Nostrand Reinhold, New York, 1995, Chapter 7.

[18] Meeker, W. Q. and Escobar, L. A., *Statistical Methods for Reliability Data*, John Wiley & Sons, New York, 1998, Chapter 13 and 21.

[19] Saunders, S. C., "Cumulative Damage Models," *Encyclopedia of Statistical Sciences,* Kotz, S. and Johnson, N.L., Eds., Vol. 2, 1982, pp. 230-233.

[20] Bauer, D. R. and Martin, J. W., Eds., *Service Life Prediction of Organic Coatings, A Systems Approach*, American Chemical Society, Washington, 1999.

[21] Meeker, W. Q., Escobar, L. A., and Chan V., "Using Accelerated Tests to Predict Service Life in Highly Variable Environments," *Service Life Prediction: Methodologies and Metrologies,* Bauer D. R. and Martin J. W. Eds., American Chemical Society, Washington, 2001.

[22] Chan V. and Meeker, W. Q., "Estimation of Degradation-Based Reliability in Outdoor Environments," *Preprint Series 2001-08*, Department of Statistics, Iowa State University, Ames, Iowa, July 2001.

[23] Jorgensen, G. J., Kim, H. M., and Wendelin, T. J., " Durability Studies of Solar Reflector Materials Exposed to Environmental Stresses," *Durability Testing of Non-metallic Materials, ASTM STP 1294*, Herling R.J., Ed., American Society for Testing and Materials, 1996.

[24] Terrenzio, L. A., Harrison, J. W., Nester, D. A., and Shiao, M. L., "Natural vs. Artificial Aging: Use of Diffusion Theory to Model Asphalt and Fiberglass-Reinforced Shingle Performance," *Proceedings of the 4th International Symposium on Roofing Technology*, National Roofing Contractors Association, Rosemont, IL, 1997, pp. 66-74.

[25] Wright J. R., "Weathering: Theoretical and Practical Aspects of Asphalt Durability," Chapter 8 in *Bituminous Materials, Vol. II: Asphalt*, Hoiberg A. J., Ed., Robert E. Krieger Publishing, Huntington, New York, 1979.

[26] Liu, M., Lunsford, K. M., Davison, R. R., Glover C. J., and Bullin, J. A., "The Kinetics of Carbonyl Formation in Asphalt," *AIChE Journal*, Vol. 42, No. 4, 1996, pp. 1069-1076.

[27] Rostler, F. S., "Fractional Composition: Analytical and Functional Significance," Chapter 6 in *Bituminous Materials, Vol. II: Asphalt*, Hoiberg A. J., Ed., Robert E. Krieger Publishing, Huntington, New York, 1979.

[28] Goldberg, L. F., Huelman, P. H., and Bridges, B. B. "A Preliminary Experimental Assessment of the Comparative Thermal Performance of Attics and Cathedral Ceilings in a Cold Climate," *ASHRAE Transactions*, vol. 105, Part 1, pp. 826-836, 1999.

[29] Wendell. A. P., Annucci, M., and Hughes, S., Quarterly Report, April 1998, Florida Energy Extension Service, Shimberg Center for Affordable Housing and Rinker School of Building Construction, University of Florida, Gainesville, Fl.

David M. Bailey,[1] Carl G. Cash,[2] and Arthur G. Davies, Jr.[3]

Update: Service Life Tests for Roofing Membranes

Reference: Bailey, D. M., Cash, C. G. and Davies, Jr., A. G., **"Update: Service Life Tests for Roofing Membranes,"** *Roofing Research and Standards Development: 5th Volume, ASTM STP 1451,* W. J. Rossiter and T. J. Wallace, Eds., ASTM International, West Conshohocken, PA, 2003.

Abstract: This research measured selected physical characteristics of twelve roofing membranes before and after oven heat and ultraviolet exposure. To make these data useful for comparing values, identical test methods and conditions were used for each membrane. We intend to compare these data with before and after exposure data obtained on companion samples exposed outdoors at three locations in the United States in an effort to determine the laboratory program that might accurately predict the performance of roofing membranes.

Keywords: APP, asphalt-glass fiber felt BUR, BUR, durability, EPDM, heat exposure, polymer modified, properties, PVC, ranking, roofing, SBS, service life, single ply, TPO, UV exposure

Introduction

We measured selected physical parameters of twelve roofing membranes before and after oven heat and ultraviolet exposure. Companion samples were exposed outdoors at three locations in the United States. Here we report our pre-exposure and post exposure heat and ultraviolet data.

Membranes Tested

Table 1 lists the twelve popular membranes that are at the core of this continuing study. With the exception of some of the polymer modified bitumen membranes that were prepared by the manufacturer, all of the multiply membranes were prepared in the laboratory under very carefully controlled conditions. The membranes were conditioned at 23°C (73°F) and 50 % relative humidity before final conditioning at the appropriate test temperature. The same personnel assembled samples for outdoor exposure at the same time.

---

[1] Civil Engineer, US Army Corps of Engineers, Construction Engineering Research Laboratory, Champaign, IL.

[2] Principal, Simpson Gumpertz & Heger Inc., Waltham, MA, Washington, DC, Los Angles, CA, and San Francisco, CA.

[3] Senior Staff Engineer – Materials/Laboratory Manager, Simpson Gumpertz & Heger Inc., Waltham, MA, Washington, DC, Los Angles, CA, and San Francisco, CA.

Table 1 – Membranes Tested

Sample	Membrane	Sample	Membrane
A	TPO thermoplastic polyolefin	G&H	APP atactic polypropylene polymer modified asphalt – 2 ply
B	PVC poly [vinyl chloride] alloy		
C&D	Asphalt-glass fiber felt BUR – 3 ply with steep asphalt	J& K	EPDM ethylene-propylene-diene terpolymer rubber
E& F	SBS styrene-butadiene-styrene polymer modified asphalt – 2 ply	L&M	PVC poly [vinyl chloride] membrane

Test Methods

Each of the membrane samples were tested in the same manner under identical conditions so that these data can be compared across all of the membrane types tested. The characterization test methods for the unexposed samples are listed in Table 2.

Table 2 – Characterization Test Methods, Unexposed Samples

Property	Method	Test Conditions
Load-strain properties Load at break Elongation Energy to first peak	D2523	25 mm (1 in.) wide strips, 100 mm (3.9 in.) jaw gap, 0.85 mm/s (2 in./min.) extension, -18°C (0°F), 23°C (73°F) and 70°C (158°F).
Cyclic fatigue resistance	D5849	-18°C (0°F), 23°C (73°F) and 70°C (158°F).
Water absorption	D570	one week in water at 60°C (140°F)
Glass transition temperature	DSC	phase transition temperature
Thermal expansion coefficient	TMA	-20°C (-4°F) to 90°C (194°F) for most samples.
Static puncture resistance	D5602	-18°C (0°F), 23°C (73°F) and 70°C (158°F)
Dynamic puncture resistance	D5635	-18°C (0°F), 23°C (73°F) and 70°C (158°F).

Legend: DSC = differential scanning calorimeter. TMA = thermal mechanical analysis.

We also ran infrared spectroscope analyses on the samples before and after oven heat and ultraviolet exposure, but these data are not included in this report because the unexposed data and the changes recorded are too hard to quantify.

Pre-exposure Test Results

The pre-exposure sample data are shown in Tables 3A [SI units] and 3B [inch-pound units]. This may be the first time you have had the opportunity to see data on this broad range of roofing membranes that is directly comparable across the different roofing types.

The breaking load of the membranes relates directly to the strength of reinforcements included. The breaking load at -18°C (0°F) exceeds 35 kilonewtons per metre width (200 force pounds per inch width) for all samples, except the EPDM samples, which are the only membranes without fiber reinforcing. The breaking load decreases as the testing temperature increases.

Table 3A – Pre-exposure Physical Characteristics of Roofing Membranes, SI Units

Sample Designation	A	B	C	D	E	F	G	H	J	K	L	M
	TPO	PVC a	BUR		SBS		APP		EPDM		PVC	
Load, kN/m												
-18°C (0°F)	37.8	42.9	38	38.5	39.2	45.3	61.9	33.8	11+	11+	37.4	35.9
23°C (73°F)	11.4	29.6	16.3	29.6	13.8	28.5	28	23.6	8.6	8.8	25.2	28.9
70°C (158°F)	3.2	23.8	3.5	6	7.8	5.6	5.1	7.4	4+	5+	9.6	17.3
Elongation , %												
-18°C (0°F)	31	37	6.2	6.3	7.6	8.8	8.3	26	278+	265+	36	32
23°C (73°F)	23	38	4.0	4.1	13	70	11	89	537	510	39	33
70°C (158°F)	5.4	40	7.9	7.5	14	21	7.7	23	282+	282+	25	32
Energy to first peak, kN/m												
-18°C (0°F)	29	30	4.6	4.2	6.5	85	10	30	74+	70+	27	22
23°C (73°F)	6.8	20	1.4	3	5.1	64	7.7	60	97	100	19	19
70°C (158°F)	0.42	18	0.54	0.91	2.6	3	0.98	4.7	24+	32+	4.9	11
Cyclic fatigue 500 cycles												
-18°C (0°F)	pass	fail	p/f	pass	pass	pass	pass	pass	pass	pass	fail	fail
23°C (73°F)	pass	pass	pass	pass	pass	pass	pass	pass	pass	pass	pass	pass
70°C (158°F)	pass	p/f	pass	pass	pass	pass	pass	pass	p/f	p/f	pass	pass
Water absorption, mass %												
	6.3	11	2.7	3.2	1.8	1.3	2.5	1.8	2.6	2.9	2.9	2.2
Glass transition temperature, °C												
	-36	-44	-21	-21	-41	-49	-34	-30	-54	-50	-42	-45
Thermal expansion x 10^{-6}/ degree C												
	75	96	204	222	376	296	361	231	83	102	86	109
Static puncture, Newtons												
-18°C (0°F)	250	250	250	250	250	250	250	250	250	250	250	250
23°C (73°F)	250	250	88	88	98	250	98	250	250	250	250	250
70°C (158°F)	250	250	67	57	35	250	47	250	250	250	250	250
Dynamic puncture, Jules												
-18°C (0°F)	3	5.6	5.6	8.1	13.1	20.6	18.1	15.6	5.6	5.6	5.6	8.1
23°C (73°F)	5.6	5.6	5.6	5.6	13.1	18.1	13.1	13.1	3	3	3	5.6
70°C (158°F)	8.1	5.6	5.6	3	3	8.1	3	5.6	3	3	5.6	5.6

The elongation and energy-to-break follow expected patterns. Like the load to break, these are ruled by the quantity and quality of the reinforcements used.

The results of the cyclic fatigue tests were mixed. The test method requires testing two samples. Sometimes one sample buckled over the moving joint while the second sample showed no distress. This is the source of the "pass/fail" or "p/f" listing on the charts.

Table 3B – Pre-exposure Physical Characteristics of Roofing Membranes, Inch-Pound Units.

Sample Designation	A TPO	B PVC a	C	D	E	F	G	H	J	K	L	M
			BUR		SBS		APP		EPDM		PVC	
Load, lbf/in.												
-18°C (0°F)	216	245	217	220	224	259	354	193	65+	65+	214	205
23°C (73°F)	65	169	93	169	79	163	160	135	49	50	144	165
70°C (158°F)	18	136	20	34	45	32	29	42	23+	29+	55	99
Elongation, %												
-18°C (0°F)	31	37	6.2	6.3	7.6	8.8	8.3	26	278+	265+	36	32
23°C (73°F)	23	38	4.0	4.1	13	70	11	89	537	510	39	33
70°C (158°F)	5.4	40	7.9	7.5	14	21	7.7	23	282+	282+	25	32
Energy to first peak, lbf/in.												
-18°C (0°F)	166	171	26	24	37	486	57	171	423+	400+	154	22
23°C (73°F)	39	114	8	17	29	366	44	343	554	571	109	22
70°C (158°F)	2	103	3	5	15	17	6	27	137+	183+	28	22
Cyclic fatigue 500 cycles												
-18°C (0°F)	pass	fail	p/f	pass	pass	pass	pass	pass	pass	pass	fail	fail
23°C (73°F)	pass	pass	pass	pass	pass	pass	pass	pass	pass	pass	pass	pass
70°C (158°F)	pass	p/f	pass	pass	pass	pass	pass	pass	p/f	p/f	pass	pass
Water absorption, mass %												
	6.3	11	2.7	3.2	1.8	1.3	2.5	1.8	2.6	2.9	2.9	2.2
Glass transition, °F												
	-33	-47	-5.8	-5.8	-42	-56	-29	-22	-65	-58	-44	-49
Thermal expansion, x 10^{-6}/°F												
x 10^{-6}/degree	135	173	367	400	677	533	650	416	149	184	155	196
Static puncture, lbf												
-18°C (0°F)	56	56	56	56	56	56	56	56	56	56	56	56
23°C (73°F)	56	56	20	20	22	56	22	56	56	56	56	56
70°C (158°F)	56	56	15	13	8	56	11	56	56	56	56	56
Dynamic puncture, ftlbf												
-18°C (0°F)	2	4	4	6	10	15	13	12	4	4	4	6
23°C (73°F)	4	4	4	4	10	13	10	10	2	2	2	4
70°C (158°F)	6	4	4	2	2	6	2	4	2	2	4	4

With the exception of Sample A, a thermoplastic polyolefin, and B, a poly[vinyl chloride] alloy with other polymers, the water absorption is in the 2 to 3 mass percent range. We suspect water was wicked into the membrane along reinforcing fibers in Sample A (6.3) and Sample B (11%).

The glass transition temperatures are the lowest for the EPDM membranes [~52°C (-62°F)]. The membranes with the highest glass transition temperature, the asphalt-glass fiber felt BUR samples [-21°C (-6°F)], have performed well in the field. The glass transition temperatures of the other samples fall between these extremes.

The thermal expansion is the average measured over a –20 to 90°C (-4 to 194°F) range for most of the samples. We measured the thermal expansion for the TPO sample over a –20 to 40°C (-4 to 104°F) range because above 40°C the expansion is non-linear. We used the –20 to 60°C (-4 to 140°F) range for the BUR and SBS samples. The asphalt containing membranes all have higher thermal expansion coefficients and lower static load resistances than the plastic or rubber membranes. A force of 250 newtons (56 pounds) is the maximum load by the static puncture resistance test.

The polymer modified asphalt samples have a greater dynamic puncture resistance than the plastic and conventional built-up membranes; these have a greater resistance than the rubber membranes.

Table 4 – Pre- and Post-exposure to Heat and Ultraviolet Data, SI Units

Sample	A - TPO	B - PVCa	C - BUR	D - BUR	E - SBS	F - SBS	G - APP	H - APP	J - EPDM	K - EPDM	L - PVC	M - PVC
Load at 23°C, kN/m												
Original	11	30	16	30	14	28	28	24	9	9	25	29
+heat	**18**	ns	**40**	ns	ns	ns	**22**	ns	ns	**8**	ns	**23**
+UV	**12**	ns	ns	**36**	ns	a	**18**	ns	**11**	**8**	ns	**25**
Elongation at 23°C, %												
Original	23	38	4	4.1	13	70	11	89	537	510	39	33
+heat	**28**	ns	**5.8**	ns	**7.3**	**88.1**	**6.3**	**68**	ns	ns	**28**	ns
+UV	**31**	**32**	ns	**4.8**	ns	a	ns	**58**	**620**	ns	ns	**39**
Energy to first peak at 23°C, kN/m												
Original	6.8	20	1.4	3	5.1	64	7.7	60	97	100	19	19
+heat	**12**	ns	**5**	ns	**2.4**	**30**	**2.6**	**47**	ns	ns	ns	ns
+UV	**15**	**16**	ns	ns	**1.9**	a	**5.5**	**38**	**134**	ns	ns	ns
Cyclic fatigue, 500 cycles												
Original	pass	pass	pass	pass	pass	pass	pass	pass	pass	pass	pass	pass
+heat	pass	pass	pass	pass	pass	pass	pass	pass	pass	pass	pass	pass
+UV	pass	pass	pass	pass	pass	a	pass	pass	pass	pass	pass	pass
Water absorption, %												
Original	6.3	11	2.7	3.1	1.8	1.3	2.5	1.8	2.6	2.9	2.9	2.2
+heat	ns	ns	ns	ns	ns	ns	ns	**2.5**	ns	ns	**3.5**	ns
+UV	ns	**14**	**9.5**	**8.3**	ns	a	**3.9**	ns	ns	ns	ns	ns
Glass transition, °C												
Original	-36	-44	-21	-21	-41	-49	-34	-30	-54	-50	-42	-45
+heat	ns	ns	**-16**	ns	**-48**	ns	**-38**	ns	ns	ns	**-44**	ns
+UV	ns	ns	ns	ns	ns	a	ns	ns	ns	ns	ns	ns
Thermal expansion, x 10^{-6}/°C												
Orignal	75	96	204	222	376	296	361	231	83	102	86	109
+heat	ns	ns	ns	ns	ns	ns	ns	ns	ns	ns	ns	ns
+UV	ns	**5.3**	ns	ns	**17.6**	a	ns	ns	**4**	**5.2**	**5.3**	ns

Static puncture resistance, N												
	250	250	67	57	35	250	47	250	250	250	250	250
+heat	ns	ns	ns	ns	ns	ns	ns	ns	ns	ns	ns	ns
+UV	Original	b	b	b	b	b	b	b	b	b	b	b
Dynamic puncture resistance, J												
Original	3	6	5.6	8.1	13.1	21	18	15.6	5.6	5.6	5.6	8.1
+heat	**8**	ns	ns	ns	ns	**18**	**13**	**13**	ns	**3**	**8.1**	**11**
+UV	**10**	**11**	**21**	**18**	31	a	**26**	**34**	**13**	**11**	**16**	**18**

a = sample destroyed during exposure; b = not performed; values in bold type are statistically different from the pre-exposure values at the 5% significance level. ns = not significant. +heat = after heat exposure; +UV = after heat exposure/

Post-exposure Test Results

We exposed one set of the prepared membrane samples to heat conditioning at 70°C (158°F) for 28 days in a forced draft oven, and a separate set of samples to cycles of 20 hours of ultraviolet light at 60°C (140°F) and 2 hours of dark condensing humidity, for a total of 1500 hours. We retested both sets of the exposed samples after conditioning the samples to room temperature.

Table 4 shows pre- and post-exposure data in SI units for tests selected to minimize the estimated standard deviation of the original data. Post-exposure data that significantly differs from the pre-exposure values is in bold type.

These data were rated by making the "best' average value in each test equivalent to 100, the "worst" equal to 0, and the others linearly proportional to these bench marks. For example, the sample with the greatest strength was rated as 100. In a similar fashion, the sample with the lowest water absorption was rated 100. After each exposure we rated the change observed. No change, or statistically insignificant changes were rated 100; the sample showing the greatest change was given a 0 change value. The other samples were assigned proportional change values. The average of the ratings for the pre-exposure and the post-exposure final values for each test, sample and exposure are shown in Table 5.

Table 5 – Pre- and Post-exposure to Heat and Ultraviolet Ratings

SAMPLE	A - TPO	B - PVCa	C - BUR	D - BUR	E -SBS	F - SBS	G -APP	H - APP	J -EPDM	K-EPDM	L - PVC	M - PVC	MEAN
Load rating													
Original	13	100	37	100	25	95	92	71	0	1	79	97	59
+heat	36	100	19	100	63	98	90	86	50	47	90	93	73
+UV	7	100	69	87	63	a	73	86	33	46	90	89	68
Elongation rating													
Original	4	6	0	0	2	12	1	16	100	95	7	5	21
+heat	28	53	0	50	2	27	3	32	100	98	23	53	39
+UV	2	30	50	26	51	a	51	8	79	98	54	27	43

Energy to first peak rating													
Original	5	19	0	2	4	63	6	59	97	100	18	18	33
+heat	37	60	0	51	42	71	40	76	99	100	59	59	58
+UV	3	51	50	51	27	a	41	65	83	100	59	59	54
Cyclic fatigue rating													
Original	100	100	100	100	100	100	100	100	100	100	100	100	100
+heat	100	100	100	100	100	100	100	100	100	100	100	100	100
+UV	100	100	100	100	100	a	100	100	100	100	100	100	100
Water absorption rating													
Original	48	0	86	80	95	100	88	95	87	84	84	96	78
+heat	74	50	93	90	98	100	94	48	94	92	65	96	83
+UV	74	44	43	58	98	a	83	98	94	92	92	96	79
Glass transition rating													
Original	45	70	0	0	61	85	39	27	100	88	64	73	54
+heat	73	85	15	50	31	93	41	64	100	94	68	87	67
+UV	73	85	50	50	81	a	70	64	100	94	82	87	76
Thermal expansion rating													
Original	100	93	57	51	0	27	5	48	97	91	96	89	63
+heat	100	97	79	76	50	64	53	74	99	96	98	95	82
+UV	100	54	79	76	0	a	53	74	50	50	62	95	63
Static puncture resistance rating													
Original	100	100	15	10	0	100	6	100	100	100	100	100	69
+heat	100	100	58	55	50	100	53	100	100	100	100	100	85
+UV	b	b	b	b	b	b	b	b	b	b	b	b	b
Dynamic puncture resistance rating													
Original	0	15	15	29	57	100	86	72	15	15	15	29	37
+heat	0	58	58	65	79	97	85	82	58	44	45	55	61
+UV	3	41	8	42	53	a	85	85	3	41	24	42	42
Mean rating													
Original	**46**	**56**	**34**	**41**	**38**	**76**	**47**	**65**	**77**	**75**	**63**	**67**	**57**
+heat	**61**	**78**	**47**	**71**	**57**	**83**	**62**	**74**	**89**	**86**	**72**	**82**	**72**
+UV	**45**	**63**	**56**	**61**	**59**	**a**	**70**	**73**	**72**	**78**	**70**	**74**	**66**

When we examine the pre-and post exposure ratings, we might predict one of the durability rankings shown in Table 6. It will be interesting to see which of these rankings are mirrored by data from the outdoor exposures. Data collections are planned after two, four, and six years of exposure at three locations in the continental United States.

Preliminary Conclusions

The EPDM, and PVC samples consistently posted the best ratings by this technique. BUR samples consistently posted the lowest ratings. This may be a function of the strength of the reinforcement within each membrane, but the EPDM samples, without reinforcement, consistently rated among the highest of this group.

It is encouraging to note that by our experience, the BUR membranes tested are giving excellent performance in the field. This could mean we can expect outstanding performance from the EPDM, SBS, and PVC membranes. It could also mean this system of rating membranes by the physical properties they possess and the change in the properties with heat and ultraviolet aging is flawed. We hope that data from our exposures will give us the answer.

Table 6 - Ranked Pre- and Post-exposure Sample Ratings

Pre-exposure		Post heat exposure		Post UV exposure	
$J_{(EPDM)}$	77	$J_{(EPDM)}$	89	$K_{(EPDM)}$	78
$F_{(SBS)}$	76	$K_{(EPDM)}$	86	$M_{(PVC)}$	74
$K_{(EPDM)}$	75	$F_{(SBS)}$	83	$H_{(APP)}$	73
$M_{(PVC)}$	67	$M_{(PVC)}$	82	$J_{(EPDM)}$	72
$H_{(APP)}$	65	$B_{(PVCa)}$	78	$G_{(APP)}$	70
$L_{(PVC)}$	63	$H_{(APP)}$	74	$L_{(PVC)}$	70
$B_{(PVCa)}$	56	$L_{(PVC)}$	72	$B_{(PVCa)}$	63
$G_{(APP)}$	47	$D_{(BUR)}$	71	$D_{(BUR)}$	61
$A_{(TPO)}$	46	$G_{(APP)}$	62	$E_{(SBS)}$	59
$D_{(BUR)}$	41	$A_{(TPO)}$	61	$C_{(BUR)}$	56
$E_{(SBS)}$	38	$E_{(SBS)}$	57	$A_{(TPO)}$	45
$C_{(BUR)}$	34	$C_{(BUR)}$	47	$F_{(SBS)}$	-
mean	58	mean	71	mean	67

Acknowledgements

The following organizations contributed resources to this study:

- Bondcoat Roofing Systems
- Carlisle Syntec Inc.
- Celotex Corporation
- Firestone Building Products Company
- GAF Materials Company
- Johns Manville International
- JPS Elastomerics Corporation
- Performance Roofing Systems
- Polyglass USA
- Sarnafil Inc.
- Simpson Gumpertz & Heger Inc.
- Siplast, Inc.
- Tamko Roofing Products, Inc.
- US Army Corps of Engineers – Engineering Research and Development Center.

The following additional organizations have promised support to continue this work:

- National Research Council of Canada
- National Roofing Contractors Association